Cosmology: A Very Short Introduction

'a fast track through the history of our endlessly fascinating Universe,
from then to now'
J. D. Barrow, Cambridge University

'In a clear and elegant style, Coles succeeds in conveying the gist of
some of the deepest concepts in Physics . . . that underlie our
understanding of the cosmos. This concise, yet up-to-date, account of
cosmic history makes for compelling reading for anyone who has ever
wondered about how our universe works.'
Carlos Frenk, University of Durham

'presents a wonderfully broad range of concepts in a clear and readable
way [and] gives a vivid picture of the confusion and excitement of
research in progress.'
P. J. E. Peebles, Princeton University

'a pleasure to read'
New Scientist

VERY SHORT INTRODUCTIONS are for anyone wanting a stimulating and accessible way in to a new subject. They are written by experts, and have been published in more than 25 languages worldwide.

The series began in 1995, and now represents a wide variety of topics in history, philosophy, religion, science, and the humanities. Over the next few years it will grow to a library of around 200 volumes – a Very Short Introduction to everything from ancient Egypt and Indian philosophy to conceptual art and cosmology.

Very Short Introductions available now:

ANCIENT PHILOSOPHY
  Julia Annas
THE ANGLO-SAXON AGE
  John Blair
ANIMAL RIGHTS   David DeGrazia
ARCHAEOLOGY   Paul Bahn
ARCHITECTURE
  Andrew Ballantyne
ARISTOTLE   Jonathan Barnes
ART HISTORY   Dana Arnold
ART THEORY   Cynthia Freeland
THE HISTORY OF
  ASTRONOMY   Michael Hoskin
ATHEISM   Julian Baggini
AUGUSTINE   Henry Chadwick
BARTHES   Jonathan Culler
THE BIBLE   John Riches
BRITISH POLITICS
  Anthony Wright
BUDDHA   Michael Carrithers
BUDDHISM   Damien Keown
CAPITALISM   James Fulcher
THE CELTS   Barry Cunliffe
CHOICE THEORY
  Michael Allingham
CHRISTIAN ART   Beth Williamson
CLASSICS   Mary Beard and
  John Henderson
CLAUSEWITZ   Michael Howard
THE COLD WAR
  Robert McMahon

CONTINENTAL PHILOSOPHY
  Simon Critchley
COSMOLOGY   Peter Coles
CRYPTOGRAPHY
  Fred Piper and Sean Murphy
DADA AND SURREALISM
  David Hopkins
DARWIN   Jonathan Howard
DEMOCRACY   Bernard Crick
DESCARTES   Tom Sorell
DRUGS   Leslie Iversen
THE EARTH   Martin Redfern
EGYPTIAN MYTHOLOGY
  Geraldine Pinch
EIGHTEENTH-CENTURY
  BRITAIN   Paul Langford
THE ELEMENTS   Philip Ball
EMOTION   Dylan Evans
EMPIRE   Stephen Howe
ENGELS   Terrell Carver
ETHICS   Simon Blackburn
THE EUROPEAN UNION
  John Pinder
EVOLUTION
  Brian and Deborah Charlesworth
FASCISM   Kevin Passmore
THE FRENCH REVOLUTION
  William Doyle
FREUD   Anthony Storr
GALILEO   Stillman Drake
GANDHI   Bhikhu Parekh

GLOBALIZATION
  Manfred Steger
HEGEL   Peter Singer
HEIDEGGER   Michael Inwood
HINDUISM   Kim Knott
HISTORY   John H. Arnold
HOBBES   Richard Tuck
HUME   A. J. Ayer
IDEOLOGY   Michael Freeden
INDIAN PHILOSOPHY
  Sue Hamilton
INTELLIGENCE   Ian J. Deary
ISLAM   Malise Ruthven
JUDAISM   Norman Solomon
JUNG   Anthony Stevens
KANT   Roger Scruton
KIERKEGAARD   Patrick Gardiner
THE KORAN   Michael Cook
LINGUISTICS   Peter Matthews
LITERARY THEORY
  Jonathan Culler
LOCKE   John Dunn
LOGIC   Graham Priest
MACHIAVELLI   Quentin Skinner
MARX   Peter Singer
MATHEMATICS   Timothy Gowers
MEDIEVAL BRITAIN
  John Gillingham and
  Ralph A. Griffiths
MODERN IRELAND
  Senia Pašeta
MOLECULES   Philip Ball
MUSIC   Nicholas Cook
NIETZSCHE   Michael Tanner
NINETEENTH-CENTURY
  BRITAIN   Christopher Harvie and
  H. C. G. Matthew
NORTHERN IRELAND
  Marc Mulholland
PAUL   E. P. Sanders
PHILOSOPHY   Edward Craig
PHILOSOPHY OF SCIENCE
  Samir Okasha

PLATO   Julia Annas
POLITICS   Kenneth Minogue
POLITICAL PHILOSOPHY
  David Miller
POSTCOLONIALISM
  Robert Young
POSTMODERNISM
  Christopher Butler
POSTSTRUCTURALISM
  Catherine Belsey
PREHISTORY   Chris Gosden
PRESOCRATIC PHILOSOPHY
  Catherine Osborne
PSYCHOLOGY   Gillian Butler and
  Freda McManus
QUANTUM THEORY
  John Polkinghorne
ROMAN BRITAIN   Peter Salway
ROUSSEAU   Robert Wokler
RUSSELL   A. C. Grayling
RUSSIAN LITERATURE
  Catriona Kelly
THE RUSSIAN REVOLUTION
  S. A. Smith
SCHIZOPHRENIA
  Chris Frith and Eve Johnstone
SCHOPENHAUER
  Christopher Janaway
SHAKESPEARE   Germaine Greer
SOCIAL AND CULTURAL
  ANTHROPOLOGY
  John Monaghan and Peter Just
SOCIOLOGY   Steve Bruce
SOCRATES   C. C. W. Taylor
SPINOZA   Roger Scruton
STUART BRITAIN   John Morrill
TERRORISM   Charles Townshend
THEOLOGY   David F. Ford
THE TUDORS   John Guy
TWENTIETH-CENTURY
  BRITAIN   Kenneth O. Morgan
WITTGENSTEIN   A. C. Grayling
WORLD MUSIC   Philip Bohlman

Available soon:

AFRICAN HISTORY
  John Parker and Richard Rathbone
ANCIENT EGYPT   Ian Shaw
THE BRAIN   Michael O'Shea
BUDDHIST ETHICS
  Damien Keown
CHAOS   Leonard Smith
CHRISTIANITY   Linda Woodhead
CITIZENSHIP   Richard Bellamy
CLASSICAL ARCHITECTURE
  Robert Tavernor
CLONING   Arlene Judith Klotzko
CONTEMPORARY ART
  Julian Stallabrass
THE CRUSADES
  Christopher Tyerman
DERRIDA   Simon Glendinning
DESIGN   John Heskett
DINOSAURS   David Norman
DREAMING   J. Allan Hobson
ECONOMICS   Partha Dasgupta
THE END OF THE WORLD
  Bill McGuire
EXISTENTIALISM   Thomas Flynn
THE FIRST WORLD WAR
  Michael Howard
FREE WILL   Thomas Pink
FUNDAMENTALISM
  Malise Ruthven
HABERMAS   Gordon Finlayson

HIEROGLYPHS
  Penelope Wilson
HIROSHIMA   B. R. Tomlinson
HUMAN EVOLUTION
  Bernard Wood
INTERNATIONAL RELATIONS
  Paul Wilkinson
JAZZ   Brian Morton
MANDELA   Tom Lodge
MEDICAL ETHICS
  Tony Hope
THE MIND   Martin Davies
MYTH   Robert Segal
NATIONALISM   Steven Grosby
PERCEPTION   Richard Gregory
PHILOSOPHY OF RELIGION
  Jack Copeland and Diane Proudfoot
PHOTOGRAPHY
  Steve Edwards
THE RAJ   Denis Judd
THE RENAISSANCE
  Jerry Brotton
RENAISSANCE ART
  Geraldine Johnson
SARTRE   Christina Howells
THE SPANISH CIVIL WAR
  Helen Graham
TRAGEDY   Adrian Poole
THE TWENTIETH CENTURY
  Martin Conway

For more information visit our web site

www.oup.co.uk/vsi

Peter Coles

COSMOLOGY

A Very Short Introduction

OXFORD
UNIVERSITY PRESS

OXFORD
UNIVERSITY PRESS

Great Clarendon Street, Oxford OX2 6DP

Oxford University Press is a department of the University of Oxford.
It furthers the University's objective of excellence in research, scholarship,
and education by publishing worldwide in

Oxford New York

Auckland Bangkok Buenos Aires Cape Town Chennai
Dar es Salaam Delhi Hong Kong Istanbul Karachi Kolkata
Kuala Lumpur Madrid Melbourne Mexico City Mumbai Nairobi
São Paulo Shanghai Taipei Tokyo Toronto

Oxford is a registered trade mark of Oxford University Press
in the UK and in certain other countries

Published in the United States
by Oxford University Press Inc., New York

© Peter Coles 2001

The moral rights of the author have been asserted
Database right Oxford University Press (maker)

First published as an Oxford University Press paperback 2001

British Library Cataloguing in Publication Data

Data available

Library of Congress Cataloging in Publication Data

Data available

ISBN 13: 978-0-19-285416-2
ISBN 10: 0-19-285416-X

5 7 9 10 8 6

Typeset by RefineCatch Ltd, Bungay, Suffolk
Printed in Great Britain by
Ashford Colour Press Ltd, Gosport, Hampshire

Preface

This book is an introduction to the ideas, methods, and results of scientific cosmology.

The subject matter of cosmology is everything that exists. The entire system of things that is the Universe encompasses the very large and the very small, the astronomical scale of stars and galaxies and the microscopic world of elementary particles. Between these limits lies a complex hierarchy of structure and pattern that results from the interplay of forces and matter. And in the midst of all this we find ourselves.

The aim of cosmology is to place all known physical phenomena within a single coherent framework. This is an ambitious goal, and significant gaps in our knowledge still remain. Nevertheless, there has been such rapid progress that many cosmologists regard this as something of a 'Golden Age'. I have taken a roughly historical path through the subject to show how it has evolved, how it has drawn together many different conceptual strands along the way, and how new avenues for exploration have opened up with improvements in technology.

It is a good time to write this kind of book. An emerging consensus about the form and distribution of matter and energy in the Universe suggests that a complete understanding of it all may be within reach. But interesting puzzles remain, and if history tells us anything it is that we should expect surprises!

Contents

List of illustrations   x

1   A brief history   1

2   Einstein and all that   12

3   First principles   27

4   The expanding Universe   39

5   The Big Bang   57

6   What's the matter with the Universe?   74

7   Cosmic structures   93

8   A theory of everything?   107

Epilogue   129

Further reading   131

Index   135

List of illustrations

1   The Babylonian God
    Marduk   3
    From *The Mythology of All Races*,
    ed. J. A. MacCulloch (Cooper Square,
    1964); see *Echoes of the Ancient Skies*,
    E. C. Krupp, p. 68.

2   Thought-experiment
    illustrating the equivalence
    principle   19
    From P. Coles, *Einstein and the Birth
    of Big Science* (Icon Books, 2000)

3   The bending of light   23
    From P. Coles, *Einstein and the Birth
    of Big Science* (Icon Books, 2000)

4   The curvature of space   25
    From P. Coles, *Einstein and the Birth
    of Big Science* (Icon Books, 2000)

5   Open, flat, and closed
    spaces in two
    dimensions   35

6   Hubble's Law   40

7   The Hubble diagram   41
    From Hubble (1929), *Proceedings
    of Nat. Acad. Sci.*, **15**, 168–173; see
    The Expanding Universe, R. C. Smith,
    p. 92

8   Redshift   46

9   The Hubble diagram
    updated   48

10 The Hubble Space
    Telescope   52
    Space Telescope Science Institute

11 Cepheids in M100   53
    Space Telescope Science Institute

12 The age of the
    Universe   55

13 The spectrum of the cosmic microwave background 60
NASA and George Smoot

14 Looking back in time 66

15 Building blocks of matter 69
Fermi National Accelerator Laboratory

16 The Friedmann models 77

17 The Coma cluster 82
National Optical Astronomy Observatories

18 Coma in X-rays 83
High Energy Astrophysics Archive Research Center

19 Gravitational lensing 84
Space Telescope Science Institute

20 The Andromeda Nebula 94
Jason Ware

21 The *Lick Map* 95
Steve Maddox

22 The 2dF galaxy redshift survey 97
Steve Maddox and the 2dF Consortium

23 The COBE ripples 100
NASA and George Smoot

24 The Hubble deep field 101
The Virgo Consortium

25 Simulation of structure formation 103
Space Telescope Science Institute

26 BOOMERANG 105
The Boomerang Collaboration

27 The flatness of space 106
The Boomerang Collaboration

28 A theory of everything 113

29 Space–time foam 116
From *300 Years of Gravity*, ed. S. W. Hawking and W. Israel (Cambridge University Press, 1987), p. 625

Chapter 1
A brief history

Cosmology is a relatively new branch of physical science. This is quite a paradoxical state of affairs, because among the questions cosmology asks are some of the most ancient that humanity has ever posed. Is the Universe infinite? Has it been around for ever? If not, how did it come into being? Will it ever come to an end? Since prehistoric times, humans have sought to build some kind of conceptual framework for answering questions about the world and their relationship to it. The first such theories or models were myths that we nowadays regard as naive or meaningless. But these primitive speculations demonstrate the importance we as a species have always attached to thinking about the Universe. Today's cosmologists use very different language and symbolism, but their motivation is largely the same as our distant ancestors. What I want to do in this chapter is briefly chart the historical development of cosmology 'the subject' and explain how some of the key ideas have evolved. I hope this will also provide a useful springboard into the other chapters in which I explore these key ideas in more detail.

The Universe in myth

Most early cosmologies are based on some form of anthropomorphism (the interpretation of something which is not human, in terms of human characteristics). Some involve the idea that the physical world is animated by wilful beings that can help or hinder mankind, others that the physical world itself is inanimate but can be manipulated by a god or gods. Either way, creation myths tend to explain the origin of the Universe in terms of entities whose motives can be understood, at least partly, by human beings.

There are many differences in creation myths around the world, but there are also some striking similarities. For one thing, their imagery often incorporates the idea of a supreme craftsman. The beauty of the natural world is thus represented as the handiwork of a skilled artisan, examples of which are found in all cultures. Another recurring image is the growth of order from chaos, mirroring the progressive organization of human society. Yet another parallel is the Universe as a biological process. The most striking examples of this occur in myths that depict the cosmos as forming from an egg or seed.

The Babylonian version of Genesis, the *Enuma Elish*, contains these elements. This myth dates from around 1450 BCE, but is probably based on much older Sumerian versions. In its account of the creation, the primordial state of disorder is identified with the sea. From the sea emerges a series of gods representing fundamental properties of the world, such as the sky, the horizon, and so on. Two of these deities, Marduk and Tiamat, fight and Tiamat the sea-goddess is killed. Marduk makes the Earth from her body.

China also furnishes interesting illustrations. One involves the giant *Pan Gu*. In this story, the cosmos began as a giant egg. The giant slept inside the egg for thousands of years before he awoke and broke free, shattering the egg in the process. Some parts of the egg (the lighter and purer bits) rose up to form the heavens while the heavier, impure parts formed the Earth. Pan Gu held up the heavens with his hands while his feet rested on the Earth. As the heavens drifted higher, the giant grew taller to keep them in contact with the Earth. Eventually Pan Gu died, but his body parts were put to good use. His left eye became the Sun,

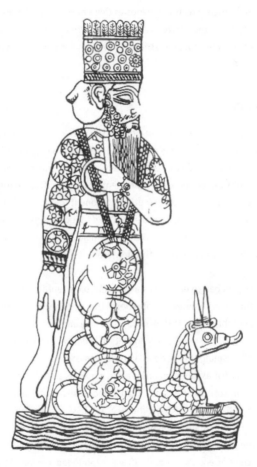

1. The Babylonian God Marduk. Marduk is credited with the imposition of cosmic order after the destruction of Tiamat, the embodiment of primordial chaos, shown here at his feet in the form of a horned dragon. Many mythologies around the world incorporate the idea that order arose from chaos, and the theme survives in some aspects of modern scientific cosmology.

his right eye the Moon. His sweat became the rain, his hair the plants of the Earth, and his bones the rocks.

There are as many creation legends as there have been cultures, and I have no space to give more examples here. Whether African, Asian, European, or American, it is striking how many formal similarities these myths display.

The Greeks

Western science has its roots in Greece. The Greeks, of course, had their own gods and myths, many of them borrowed from neighbouring cultures. But alongside these more traditional elements they began to establish a system of principles for scientific enquiry. The identification of cause and effect, still an essential component of scientific theories, was down to the Greeks. They also realized that descriptions and explanations of observed phenomena could be phrased in mathematical or geometrical rather than anthropomorphic terms.

Cosmology began to emerge as a recognizable scientific discipline within the overall framework of rational thought constructed by the Greeks, notably through Thales (625–547 BCE) and Anaximander (610–540 BCE). The word cosmology itself is derived from the Greek 'cosmos', meaning the world as an ordered system or whole. The emphasis is just as much on order as on wholeness, for in Greek the opposite of 'cosmos' is 'chaos'. The Pythagoreans of the sixth century BCE regarded numbers and geometry as the basis of all natural things. The advent of mathematical reasoning, and the idea that one can learn about the physical world using logic and reason marked the beginning of the scientific era. Plato (427–348 BCE) expounded a complete account of the creation of the Universe, in which a divine Demiurge creates, in the physical world, imperfect representations of the structures of pure being that exist only in the world of ideas. The physical world is subject to change, whereas the world of ideas is eternal and immutable.

Aristotle (384–322 BCE), a pupil of Plato, built on these ideas to present a picture of the world in which the distant stars and planets execute perfect circular motions, circles being a manifestation of 'divine' geometry. Aristotle's Universe is a sphere centred on the Earth. The part of this sphere that extends as far as the Moon is the domain of change, the imperfect reality of Plato, but beyond this the heavenly bodies execute their idealized circular motions. This view of the Universe was to dominate Western European thought throughout the Middle Ages, but its perfect circular motions did not match the growing quantities of astronomical data being gathered by the Greeks from the astronomical archives made by the Babylonians and Egyptians. Although Aristotle had emphasized the possibility of learning about the Universe by observation as well as pure thought, it was not until Ptolemy's *Almagest*, compiled in the second century CE, that a complete mathematical model for the Universe was assembled that agreed with all the data available.

The Renaissance

Much of the knowledge acquired by the Greeks was lost to Christian culture during the dark ages, but it survived in the Islamic world. As a result, cosmological thinking during the Middle Ages of Europe was rather restricted. Thomas Aquinas (1225–74) seized on Aristotle's ideas, which were available in Latin translation at the time while the *Almagest* was not, to forge a synthesis of pagan cosmology with Christian theology which was to dominate Western thought until the sixteenth and seventeenth centuries.

The dismantling of the Aristotelian world-view is usually credited to Nicolaus Copernicus (1473–1543). Ptolemy's *Almagest* was a complete theory, but it involved applying a different mathematical formula for the motion of each planet and therefore did not really represent an overall unifying system. In a sense, it described the phenomena of heavenly motion but did not explain them. Copernicus wanted to derive a single universal theory that treated everything on the same footing. He achieved this only partially, but did succeed in displacing the Earth from the centre of the scheme of things. It was not until Johannes Kepler (1571–1630) came along that a completely successful demolition of the Aristotelian system was achieved. Driven by the need to explain the highly accurate observations of planetary motion made by Tycho Brahe (1546–1601), Kepler replaced Aristotle's divine circular orbits with ellipses.

The next great development on the road to modern cosmological thinking was the arrival on the scene of Isaac Newton (1642–1727). Newton was able to show, in his monumental *Principia* (1687), that the elliptical motions devised by Kepler were the natural outcome of a universal law of gravitation. Newton therefore re-established a kind of Platonic level of reality, the idealized world of universal laws of motion. The Universe, in Newton's picture, behaves as a giant machine, enacting the regular motions demanded by the divine Creator and both time and space are absolute manifestations of an internal and omnipresent God.

Newton's ideas dominated scientific thinking until the beginning of the twentieth century, but by the nineteenth century the cosmic machine had developed imperfections. The mechanistic world-view had emerged alongside the first stirrings of technology. During the subsequent Industrial Revolution, scientists had become preoccupied with the theory of engines and heat. These laws of thermodynamics had shown that no engine could work perfectly for ever without running down. In this time there arose a widespread belief in the 'Heat Death of the Universe', the idea that the cosmos as a whole would eventually fizzle out just as a bouncing ball gradually dissipates its energy and comes to rest.

Towards the modern era

Another spanner was thrown into the works of Newton's cosmic engine by Olbers (1758–1840), who formulated in 1826 a paradox that still bears his name although it was discussed by many before him, including Kepler. *Olbers' Paradox* emerges from considering why the night sky is dark. In an infinite and unchanging Universe, every line of sight from an observer should hit a star, in much the same way as a line of sight through an infinite forest will eventually hit a tree. The consequence of this is that the night sky should be as bright as a typical star. The observed darkness at night is sufficient to prove the Universe cannot be both infinite and eternal.

Whether the Universe is infinite or not, the part of it accessible to rational explanation has steadily increased. For Aristotle, the Moon's orbit (a mere 400,000 km) marked a fundamental barrier, beyond which the human mind could not reach. To Copernicus and Kepler the limit was the edge of the Solar System (billions of kilometres away). By the eighteenth and nineteenth centuries, it was being suggested that the Milky Way, a structure now known to be at least a billion times larger than the Solar System, was the entire Universe. This suggestion had a rival, the idea that many strange spiral 'nebulae' discovered scattered across the sky were objects very similar to our Milky Way but seen at immense distances. These objects would come to be called galaxies. A 'Great Debate' took place in the early years of the twentieth century between these two opposing ideas, which I will discuss in Chapter 4. Thanks largely to Edwin Hubble (1889–1953), it is now known that the Milky Way is indeed only one of hundreds of billions of similar galaxies.

The modern era of cosmology began in the early years of the twentieth century, with a complete rewrite of the laws of nature. Albert Einstein (1879–1955) introduced the principle of relativity in 1905 and thereby demolished Newton's conception of space and time. Later, his general theory of relativity also supplanted Newton's law of universal gravitation. The first great works on relativistic cosmology by Friedmann (1888–1925), Lemaître (1894–1966), and de Sitter (1872–1934) formulated a new and complex language for the mathematical description of the Universe. Einstein's theory plays such a fundamental conceptual role in modern cosmology that I will devote much of the next chapter to it.

But while these conceptual developments paved the way, the final steps towards the modern era were taken not by theoretical physicists, but by observational astronomers. In 1929, Edwin Hubble, who had only recently shown that the Universe contained many galaxies like the Milky Way, published the observations that led to the realization that our Universe is expanding. Finally, in 1965, Penzias and Wilson discovered the cosmic microwave background, proof (or as near to proof as you're likely to see) that our Universe began in a primordial fireball – the Big Bang.

Cosmology today

The modern era of scientific cosmology began with Einstein's general theory of relativity, published in 1915, which made possible a consistent mathematical description of the entire Universe. According to Einstein's theory, the properties of matter and motion are related to deformations of space and time. The importance of this for cosmology is that space and time are no longer thought of as absolute and independent of material bodies, but as participants in the evolution of the Universe. General relativity allows us to understand not the origin of the cosmos *in* space and time, but the origin *of* space and time themselves.

Einstein's theory forms the basis of the modern Big Bang model, which has emerged as the best available description of the expanding Universe. According to this model, space, time, matter, and energy all came into existence as a primordial fireball of matter and radiation at extremes of temperature and density about 15 billion years ago. A few seconds after the beginning, the temperature had decreased to a mere 10 billion degrees and nuclear reactions began to make the atoms from which we are all made. After about 300,000 years the temperature had fallen to a few thousand degrees, releasing the radiation we now observe as the cosmic microwave background. As this explosion expanded, carrying space and time with it, the Universe cooled and rarefied. Stars and galaxies formed by condensing out of the expanding cloud of gas and radiation. Our present-day Universe contains the ashes and smoke left over from the Big Bang.

Chapter 5 describes the Big Bang theory in more detail. Most cosmologists accept it as being essentially correct, as far as it goes. It explains most of the things we know about the bulk properties of the Universe, and can account for most relevant cosmological observations. But it is important to realize that the Big Bang is not complete. Much of modern cosmological research is driven by the desire to fill the gaps in this otherwise compelling framework.

For one thing, Einstein's theory itself breaks down at the very beginning of the Universe. The Big Bang is an example of what relativity theorists call a *singularity*, a point where the mathematics falls to pieces and measurable quantities become infinite. While we know how the Universe is expected to evolve from a given stage, the singularity makes it impossible to know from first principles what the Universe should look like in the beginning. We therefore have to piece this together using observations rather than pure thought, like archaeologists trying to reconstruct a city from ruins. Modern-day cosmologists are therefore collecting huge quantities of detailed data so that they can try to piece it all together to make a picture of how the Universe began.

Technological developments over the last twenty years have accelerated progress in observational cosmology to a remarkable extent, and we are truly in a 'Golden Age' of cosmic discovery. Observational cosmology now includes the construction of huge maps of the distribution of galaxies in space, showing the remarkable large-scale structure of filaments and sheets. These surveys are complemented by deep observations being made with, for example, the Hubble Space Telescope. The Hubble Deep Field is such a long exposure that it can see galaxies at distances so huge that it has taken light much of the age of the Universe to reach us from them. Using observations like this we can see cosmic history unfolding. For example, microwave astronomers are now able to make pictures of the structure of the early Universe by observing ripples in the cosmic microwave background produced in the primordial fireball. Planned satellite experiments, such as MAP and the Planck Surveyor, will probe these ripples in more detail over the next few years and the results they produce should plug many of the gaps in our understanding of how the Universe is put together.

Astronomical observations can be used to measure the rate of cosmic expansion, how this is changing with time, and also to probe the geometry of space by applying the principles of triangulation on an enormous scale. In Einstein's theory, light rays do not necessarily travel in straight lines because space distorts in response to the gravity produced by massive bodies. Over cosmological distances, this effect can close the whole of space-time back on itself (like the surface of a sphere) causing the paths of parallel light rays to converge. It could also produce an 'open' Universe in which light rays diverge. Poised in between these two alternatives is the 'normal' idea of flat space in which Euclid's laws of geometry apply. Which of these alternatives is correct depends upon the total cosmic density of matter and energy, which the Big Bang theory cannot itself predict.

The Big Bang theory underwent a major theoretical overhaul in the early 1980s, when particle physicists took up cosmology as a way of trying to understand the properties of matter at the extremely high energies their particle accelerators couldn't reach. These theorists realized that the early Universe would be expected to undergo a series of dramatic transformations known as phase transitions, during which its expansion would accelerate by an enormous factor in a tiny fraction of a second. Such a period of 'inflation' is expected to flatten out the curvature of space leading to a definite prediction that the Universe should be flat. This seems to be consistent with the cosmic surveys mentioned above. Recent suggestions that the expansion of the Universe may be accelerating even now, suggest the existence of a mysterious dark energy that is perhaps some relic of the earlier inflationary phase.

Cosmologists have also applied modern supercomputers to the business of trying to understand the condensation of clumps of cosmic material into stars and galaxies as the Universe expands and cools. These calculations have suggested that this process requires the existence of huge concentrations of exotic material, dense enough to assist the growth of structure, yet producing no starlight. This invisible stuff is called *dark matter*. Computer predictions of the structure formed are in close agreement with the huge maps being made by the observers lending further support to the Big Bang theory.

The interplay between these new theoretical ideas and new high-quality observational data has catapulted cosmology from the purely theoretical domain and into the field of rigorous experimental science. This process began at the beginning of the twentieth century, with the work of Albert Einstein.

Chapter 2
Einstein and all that

We are all aware of the effects of gravity. Objects fall to Earth when we drop them. It's harder to run uphill than down. To a physicist, however, there is much more to gravity than its effects on our everyday lives. For one thing, the larger the scale of things being considered, the more important gravity becomes. Gravity pulls the Earth around the Sun, and the Moon around the Earth, and it causes tides. On the scale of things relevant to astronomy, gravity is the prime mover. So if you want to understand the Universe as a whole, you have to understand gravity.

Universal gravitation

Gravity is one of the fundamental forces of nature. It represents the universal tendency of all matter to attract all other matter. There are in fact four fundamental forces (gravity, electromagnetism, and the 'weak' and 'strong' nuclear forces). The universality of gravity sets it apart from, for example, the electrical forces between charged bodies. Electrical charges can be of two different kinds, positive or negative. While electrical forces can lead either to attraction (between unlike charges) or repulsion (between like charges), gravity is always attractive. That is why it is so important for cosmology.

In many ways, the force of gravity is extremely weak. Most material bodies are held together by electrical forces between atoms which are many orders of magnitude stronger than the gravitational forces between them. But, despite its weakness, gravity is the driving force in astronomical situations because astronomical bodies, with very few exceptions, always contain exactly the same amount of positive and negative charge and therefore never exert forces of an electrical nature on each other.

One of the first great achievements of theoretical physics was Isaac Newton's theory of universal gravitation, which unified what, at the time, seemed to be many disparate physical phenomena. Newton's theory of mechanics is encoded in three simple laws:

1. Every body continues in a state of rest or uniform motion in a straight line unless it is compelled to change that state by forces impressed upon it.
2. Rate of change of momentum is proportional to the impressed force, and is in the direction in which this force acts.
3. To every action, there is always opposed an equal reaction.

These three laws of motion are general, applying just as accurately to the behaviour of balls on a billiard table as to the motion of the heavenly bodies. All that Newton needed to do was to figure out how to describe the force of gravity. Newton realized that a body orbiting in a circle, like the Moon going around the Earth, is experiencing a force in the direction of the centre of motion (just as a weight tied to the end of a piece of string does when it is twirled around one's head). Gravity could cause this motion in the same way as it could cause apples to fall to Earth from trees. In both these situations, the force has to be towards the centre of the Earth. Newton realized that the right form of mathematical equation was an 'inverse-square' law: 'the attractive force between any two bodies depends on the product of the masses of the bodies and upon the square of the distance between them.'

It was a triumph of Newton's theory, based on the inverse-square law of universal gravitation, that it could explain the laws of planetary motion obtained by Johannes Kepler more than a century earlier. So spectacular was this success that the idea of a Universe guided by Newton's laws of motion was to dominate scientific thinking for more than two centuries, until the arrival on the scene of Albert Einstein.

The Einstein revolution

Albert Einstein was born in Ulm (Germany) on 14 March 1879, but his family soon moved to Munich, where he spent his school years. The young Einstein was not a particularly good student, and in 1894 he dropped out of school entirely when his family moved to Italy. After failing the entrance examination once, he was eventually admitted to the Swiss Institute of Technology in Zurich in 1896. Although he did fairly well as a student in Zurich, he was unable to get a job in any Swiss university, as he was held to be extremely lazy. He left academia to work in the Patent Office at Bern in 1902. This gave him a good wage and, since the tasks given to a junior patent clerk were not exactly onerous, it also gave him plenty of spare time to think about physics.

Einstein's special theory of relativity was published in 1905. It stands as one of the greatest intellectual achievements in the history of human thought. It is made even more remarkable by the fact that Einstein was still working as a patent clerk at the time, and was doing physics as a particularly demanding hobby. What's more, he also published seminal works that year on the photoelectric effect (which was to inspire many developments in quantum theory) and on the phenomenon of Brownian motion (the jiggling of microscopic particles as they are buffeted by atomic collisions). But the reason why the special theory of relativity stands head and shoulders above his own work of this time, and that of his colleagues in the world of mainstream physics, is that Einstein managed to break away completely from the concept of time as an absolute property that marches on at the same rate for everyone and everything. This idea is built into the Newtonian picture of the world, and most of us regard it as being so obviously true that it does not bear discussion. It takes a genius to break down conceptual barriers of such magnitude.

The idea of relativity did not originate with Einstein. Galileo had articulated the basic principle nearly three centuries earlier. Galileo claimed that only relative motion matters, so there could be no such thing as absolute motion. He argued that if you were travelling in a boat at constant speed on a smooth lake, then there would be no experiment that you could do in a sealed cabin on the boat that would indicate to you that you were moving at all. Of course, not much was known about physics in Galileo's time, so the kinds of experiment he could envisage were rather limited.

Einstein's version of the principle of relativity simply turned it into the statement that all laws of nature have to be exactly the same for all observers in relative motion. In particular, Einstein decided that this principle must apply to the theory of electromagnetism, constructed by James Clerk Maxwell, which describes amongst other things the forces between charged bodies mentioned above. One of the consequences of Maxwell's theory is that the speed of light (in vacuum) appears as a universal constant (usually given the symbol 'c'). Taking the principle of relativity seriously means that all observers have to measure the same value of c, whatever their state of motion. This seems straightforward enough, but the consequences are nothing short of revolutionary.

Einstein decided to ask himself specific questions about what would be observed in particular kinds of experiments involving the exchange of light signals. He worked a great deal with *gedanken* (thought) experiments of this kind. For example, imagine there is a flash bulb in the centre of a railway carriage moving along a track. At each end of the carriage there is a clock, so that when the flash illuminates it we can see the time. If the flash goes off, then the light signal reaches both ends of the carriage simultaneously, from the point of view of passengers sitting in the carriage. The same time is seen on each clock.

Now picture what happens from the point of view of an observer at rest who is watching the train from the track. The light flash travels with the same speed in our reference frame as it did for the passengers. But the passengers at the back of the carriage are moving into the signal, while those at the front are moving away from it. This observer therefore sees the clock at the back of the train light up before the clock at the front does. But when the clock at the front does light up, it reads the same time as the clock at the back did! This observer has to conclude that something is wrong with the clocks on the train.

This example demonstrates that the concept of simultaneity is relative. The arrivals of the two light flashes are simultaneous in the frame of the carriage, but occur at different times in the frame of the track. Other examples of strange relativistic phenomena include time dilation (moving clocks appear to run slow) and length contraction (moving rulers appear shorter). These are all consequences of the assumption that the speed of light must be the same as measured by all observers. Of course, the examples given above are a little unrealistic. In order to show noticeable effects, the velocities concerned must be a sizeable fraction of c. Such speeds are unlikely to be reached in railway carriages. Nevertheless, experiments have been done that show that time dilation effects are real. The decay rate of radioactive particles is much slower when they are moving at high velocities because their internal clock runs slowly.

Special relativity also spawned the most famous equation in all of physics: $E = mc^2$, expressing the equivalence between matter and energy. This has also been tested experimentally; amongst other things it is the principle behind both atomic and chemical explosives.

Remarkable though the special theory undoubtedly is, it is seriously incomplete because it deals only with bodies moving with constant velocity with respect to each other. Even chapter 1 of the laws of nature, written by Newton, had been built around the causes and consequences of velocities that change with time. Newton's second law is about the rate of change of momentum of an object, which in layman's terms is its acceleration. Special relativity is restricted to so-called inertial motions, i.e. the motions of particles that are not acted upon by any external forces. This means that special relativity cannot describe accelerated motion of any kind and, in particular, cannot describe motion under the influence of gravity.

The equivalence principle

Einstein had deep insights into how to incorporate gravitation into relativity theory. For a start, consider Newton's theory of gravity. In this theory, the force on a particle of mass m due to another particle of mass M depends on the product of these masses and the square of the distance between the particles. According to Newton's laws of motion, this induces an acceleration in the first particle given by $F = ma$. The m in this equation is called the inertial mass of the particle, and it determines the particle's resistance to being accelerated. In the inverse-square law of gravity, however, the mass m measures the reaction of one particle to the gravitational force produced by the other particle. It is therefore called the passive gravitational mass. But Newton's third law of motion also states that if body A exerts a force on body B then body B exerts a force on body A which is equal and opposite. This means that m must also be the active gravitational mass (if you like, the gravitational charge) produced by the particle. In Newton's theory, all three of these masses – the inertial mass, the active and passive gravitational masses – are equivalent. But there seems to be no reason, on the face of it, why this should be the case. Couldn't they be different?

Einstein decided that this equivalence must be the consequence of a deeper principle called the principle of equivalence. In his own words, this means that 'all local, freely-falling laboratories are equivalent for the performance of all physical experiments'. What this means is essentially that one can do away with gravity as a separate force of nature and regard it instead as a consequence of moving between accelerated frames of reference.

To see how this is possible, imagine a lift equipped with a physics laboratory. If the lift is at rest on the ground floor, experiments will reveal the presence of gravity to the occupants. For example, if we attach a weight on a spring to the ceiling of the lift, the weight will extend the spring downwards. Next, imagine that we take the lift to the top of a building and let it fall freely. Inside the freely falling lift there is no perceptible gravity. The spring does not extend, as the weight is always falling at the same rate as the rest of the lift, even though the lift's speed might be changing. This is what would happen if we took the lift out into space, far away from the gravitational field of the Earth. The absence of gravity therefore looks very much like the state of free-fall in response to a gravitational force. Moreover, imagine that our lift was actually in space (and out of gravity's reach), but there was a rocket attached to it. Firing the rocket would make the lift accelerate. There is no up or down in free space, but let us assume that the rocket is attached so that the lift would accelerate in the opposite direction from before, i.e. in the direction of the ceiling.

What happens to the spring? The answer is that the acceleration makes the weight move in the reverse direction relative to the lift, thus extending the spring towards the floor. (This is like what happens when a car suddenly accelerates – the passenger's head is flung backwards.) But this is just like what happened when there was a gravitational field pulling the spring down. If the lift carried on accelerating, the spring would remain extended, just as if it were not accelerating but placed in a gravitational field. Einstein's idea was that these situations do not merely appear similar: they are completely indistinguishable. Any experiment performed in an accelerated lift in space would give exactly the same results as one performed in a lift upon which gravity is acting. To complete the picture, now consider a lift placed inside a region

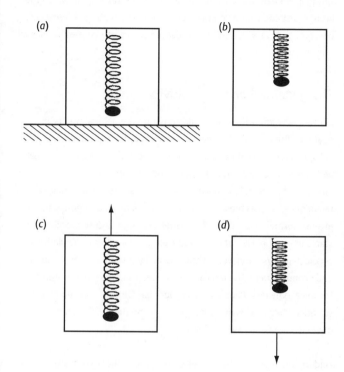

2. Thought-experiment illustrating the equivalence principle. A weight is attached to a spring, which is attached to the ceiling of a lift. In (a) the lift is stationary, but a gravitational force acts downwards; the spring is extended by the weight. In (b) the lift is in deep space, away from any sources of gravity, and is not accelerated; the spring does not extend. In (c) there is no gravitational field, but the lift is accelerated upwards by a rocket; the spring is extended. The acceleration in (c) produces the same effect as the gravitational force in (a). In (d) the lift is freely falling in a gravitational field, accelerating downwards so no gravity is felt inside; the spring does not extend because in this case the weight is weightless and the situation is equivalent to (b).

where gravity is acting, but which is allowed to fall freely in the gravitational field. Everything inside becomes weightless, and the spring is not extended. This is equivalent to the situation in which the lift is at rest and where no gravitational forces are acting. A freely falling observer has every reason to consider himself to be in a state of inertial motion.

The general theory of relativity

Einstein now knew how he should construct the general theory of relativity. But it would take him another ten years to produce the theory in its final form. What he had to find was a set of laws that could deal with any form of accelerated motion and any form of gravitational effect. To do this he had to learn about sophisticated mathematical techniques, such as tensor analysis and Riemannian geometry, and to invent a formalism that was truly general enough to describe all possible states of motion. He got there, but clearly it wasn't easy. While his classic papers of 1905 were characterized by brilliant clarity of thought and economy of mathematical calculation, his later work is mired in technical difficulty. People have argued that Einstein grew up as a scientist while he was developing the general theory. If so, it was obviously a difficult process for him.

Understanding the technicalities of the general theory of relativity is a truly daunting task. Even on a conceptual level, the theory is difficult to grasp. The relativity of time embodied in the special theory is present in the general theory, but there are additional effects of time dilation and length contraction due to gravitational effects. And the problems don't end with time! In the special theory, space at least is well behaved. In the general theory, even this goes out of the window. Space is curved.

The curvature of space

The idea that space could be warped is so difficult to grasp that even physicists don't really try to visualize such a thing. Our understanding of the geometrical properties of the natural world is based on the achievements of generations of Greek mathematicians, notably the formalized system of Euclid – Pythagoras' theorem, parallel lines never meeting, the sum of the angles of a triangle adding up to 180 degrees, and so on. All of these rules find their place in the canon of Euclidean geometry. But these laws and theorems are not just abstract mathematics. We know from everyday experience that they describe the properties of the physical world extremely well. Euclid's laws are used every day by architects, surveyors, designers, and cartographers – anyone, in fact, who is concerned with the properties of shape, and the positioning of objects in space. Geometry is real.

It seems self-evident, therefore, that these properties of space that we have grown up with should apply beyond the confines of our buildings and the lands we survey. They should apply to the Universe as a whole. Euclid's laws must be built into the fabric of the world. Or must they? Although Euclid's laws are mathematically elegant and logically compelling, they are not the only set of rules that can be used to build a system of geometry. Mathematicians of the nineteenth century, such as Gauss and Riemann, realized that Euclid's laws represent only a special case of geometry wherein space is flat. Different systems can be constructed in which these laws are violated.

Consider, for example, a triangle drawn on a flat sheet of paper. Euclid's theorems apply here, so the sum of the internal angles of this triangle must be 180 degrees (equivalent to two right-angles). But now think about what happens if you draw a triangle on a sphere instead. It is quite possible to draw a triangle on a sphere that has three right angles in it. For example, draw one point at the 'north pole' and two on the 'equator' separated by one quarter of the circumference. These three points form a triangle with three right angles that violates Euclidean geometry.

Thinking this way works fine for two-dimensional geometry, but our world has three dimensions of space. Imagining a three-dimensional curved surface is much more difficult. But in any case it is probably a mistake to think of 'space' at all. After all, one can't measure space. What one can measure are distances between objects located in space using rulers or, more realistically in an astronomical context, light beams. Thinking of space as a flat or curved piece of paper encourages one to think of it as a tangible thing in itself, rather than simply as where the tangible things are not. Einstein always tried to avoid dealing with entities such as 'space' whose category of existence was unclear. He preferred to reason instead about what an observer could actually expect to measure with a given experiment.

Following this lead, we can ask what kind of path light rays follow according to the general theory of relativity. In Euclidean geometry, light travels on straight lines. We can take the straightness of light paths to mean essentially the same thing as the flatness of space. In special relativity, light also travels on straight lines, so space is flat in this view of the world too. But remember that the general theory applies to accelerated motion, or motion in the presence of gravitational effects. What happens to light in this case?

Let us go back to the thought experiment involving the lift. Instead of a spring with a weight on the end, the lift is now equipped with a laser beam that shines from side to side. The lift is in deep space, far from any sources of gravity. If the lift is stationary, or moving with constant velocity, then the light beam hits the side of the lift exactly opposite to the laser device that produces it. This is the prediction of the special theory of relativity. But now imagine the lift has a rocket which switches on and accelerates it upwards. An observer outside the lift who is at rest

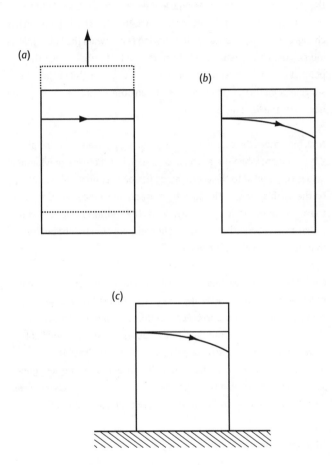

3. The bending of light. In (*a*), our lift is accelerating upwards, as in Fig. 2(*c*). Viewed from outside, a laser beam follows a straight line. In (*b*), viewed inside the lift, the light beam appears to curve downwards. The effect in a stationary lift situated in a gravitational field is the same, as we see in (*c*).

sees the lift accelerate away, but if he could see the laser beam from outside it would still be straight. On the other hand, a physicist inside the lift notices something strange. In the short time it takes light to travel from one side of the lift to the other, the lift's state of motion has changed. It has accelerated so it is moving faster when the light ends its journey than it was when the light started out. This means that the point at which the laser beam hits the other wall is slightly below the starting point on the other side. Seen by an observer inside, acceleration has 'bent' the light ray downwards.

Now remember the case of the spring and the equivalence principle. What happens when there is no acceleration but there is a gravitational field is very similar to the accelerated lift. Consider now a lift standing on the Earth's surface. The light ray must do much the same thing as in the accelerating lift: it bends downward. The conclusion we are led to is that gravity bends light. And if light paths are not straight but bent, then space is not flat but curved.

One of the reasons we find curved space hard to understand is that we don't observe it in everyday life. This is because gravity is so weak in commonplace circumstances. Even on the scale of our Solar System, gravity is so weak that the curvature it causes is negligible and light travels in lines that are so nearly straight that we can't tell the difference. In these situations, Newton's laws of motion are very good approximations to what happens. There are cases, however, where we must be prepared to deal with strong gravity and all that implies.

Black holes and the Universe

One example where Newton's gravity breaks down is when a very large amount of matter is concentrated in a very small region of space. When this happens the action of gravity is so strong, and space so warped, that light is not merely bent but is trapped. Such an object is a black hole.

(a)

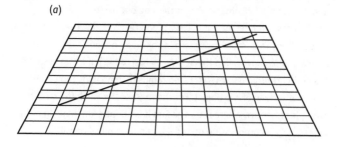

(b)

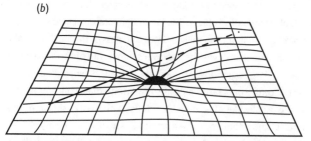

4. The curvature of space. In the absence of a source of gravitation, light travels in a straight line. If a massive object is placed near the light path, the distortion of space produces a bent light ray.

The idea that black holes might exist in nature dates back to John Michell, an English clergyman, in 1783, and was also discussed by Laplace. Such objects are, however, most commonly associated with Einstein's theory of general relativity. Indeed, one of the first mathematical solutions of Einstein's equations obtained, describes such an object. The famous 'Schwarzschild' solution was obtained only a year after the publication of Einstein's theory in 1916 by Karl Schwarzschild, who died soon after on the eastern front in the First World War. The solution corresponds to a spherically symmetric distribution of matter, and it was originally intended that this could form the basis of a mathematical model for a star. It was soon realized, however, that for an object of any mass the Schwarzschild solution implied the existence of a critical radius (now called the Schwarzschild radius). If a massive object lies entirely within its Schwarzschild radius then no light can escape from the surface of the object. For the mass of the Earth the critical radius is only 1 cm, whereas for the Sun it is about 3 km. Making black holes involves compressing material to a phenomenal density.

Since the pioneering work of Schwarzschild, research on black holes has been intense. Although there is as yet no truly watertight direct evidence for the existence of black holes in nature, there is a mountain of circumstantial evidence suggesting they might be lurking in many kinds of astronomical object. The intense gravitational field surrounding a black hole of about 100 million times the mass of the Sun is thought to be the engine that drives the enormous luminosity of certain types of galaxies. More recent observational studies of the dynamics of stars near the centre of galaxies indicate very strong mass concentrations that are usually identified with black holes with masses similar to this figure. It is now thought to be a serious possibility that nearly all galaxies have a black hole in their core. Black holes of much smaller mass may be formed at the end of the life of a star, when its energy source fails and it collapses in on itself.

There is a great deal of interest nowadays in black holes, but they are not central to the development of cosmology so I shall not discuss them further in this book. Instead, in the next chapter I'll discuss the role Einstein's theory has played in understanding the behaviour of the Universe as a whole.

Chapter 3
First principles

Einstein published his general theory of relativity in 1915. Almost immediately he sought to exploit this new theoretical framework to explain the large-scale behaviour of the entire cosmos. He was handicapped in his pursuit of this goal by the lack of information available to him about what it was that he was attempting to explain. What was the Universe really like? Einstein's knowledge of astronomy was scanty, but he needed to know the answers to some fundamental questions before he could proceed. He knew that pure thought alone could not tell him what the Universe should look like and how it should behave. Observations and guesswork would have to guide him.

Simplicity and symmetry

There is no doubt that the general theory furnishes an elegant conceptual framework, as I have tried to explain using thought experiments and pictures. The harsh truth, however, is that it involves some of the most difficult mathematics ever applied to a description of nature. To give some idea of the complexity involved, it is useful to compare Einstein's theory with the older Newtonian approach.

In Newton's theory of motion there is basically one mathematical equation to solve. This equation is '$F = ma$', and it relates the force F on an object to the acceleration a of that object. It sounds simple enough, but in practice it can be overwhelmingly complicated to describe gravity using this approach. The reason is that every piece of matter in the Universe exerts a gravitational force on every other. It's relatively easy to apply this idea to the motion of two interacting bodies, such as the Earth and the Sun, but if you start to add more bodies then things get very sticky. Indeed, while there is an exact mathematical solution to Newton's theory for two orbiting bodies, there is no known general solution for any situation more complicated than this. Not even three bodies. Applying Newton's theory to systems comprising large collections of gravitating objects is very difficult and usually requires the use of powerful computers to understand what is happening. The only exception is when the system involves some simplifying symmetry, such as a sphere, or has components that are distributed uniformly through space.

Newton's gravity is hard enough to apply in realistic situations, but Einstein's theory is an absolute nightmare. For one thing, instead of Newton's one equation, Einstein has no less than ten, which must all be solved simultaneously. And each separate equation is much more complicated than Newton's simple inverse-square law. Because of the equivalence between mass and energy given by $E = mc^2$, all forms of energy gravitate. The gravitational field produced by a body is itself a form of energy, and it also therefore gravitates. This kind of chicken-and-egg problem is called 'non-linearity' by physicists, and it often leads to unmanageable mathematical complexity when it comes to solving the equations. This is the case for general relativity. Exact mathematical solutions of Einstein's equations are very few and far between. Even with special symmetry the theory poses grand challenges for mathematicians and computers alike.

Einstein knew that his equations were hard to solve, and that he would not be able to make much progress unless he assumed the Universe had some simplifying symmetry or uniformity. In 1915 relatively little was known for sure about the way in which the contents of the Universe were distributed. Many astronomers felt that the Milky Way was an 'Island Universe'; others believed that it was just one of many such objects scattered more or less uniformly throughout space. The latter possibility appealed most to Einstein. The Milky Way is an ugly slab of gas, dust, and stars that would be very difficult to describe properly if it were the whole Universe. The second option was better in that it allowed a rough-and-ready description in which the Milky Way and other galaxies were the fine details in a largely smooth distribution of material. Einstein also had philosophical reasons for preferring large-scale smoothness, stemming from an idea called Mach's principle. If the Universe were the same everywhere he could set his cosmological theory on a solid footing by allowing the distribution of matter to define a special reference frame that would help him deal with the effects of gravity.

So, with precious little observational evidence to go on, Einstein decided that he would simplify the Universe he described by making it homogeneous (i.e. the same in every place); at least on scales much larger than the observed lumpy bits (i.e. the galaxies). He also assumed the Universe to be isotropic (i.e. looking the same in every direction). These twin assumptions together form the *Cosmological Principle*.

The Cosmological Principle

The twin assumptions of homogeneity and isotropy are related but not equivalent. Isotropy does not necessarily imply homogeneity without the additional assumption that the observer is not in a special place. One would observe isotropy in any spherically symmetric distribution of matter, but only if one were in the middle. A circular carpet with a pattern consisting of a series of concentric rings would look isotropic only to an observer standing in the centre of the pattern. The principle that we do not live in a special place in the Universe is called the *Copernican Principle*, an indication of the debt that modern cosmology owes to history. Observed isotropy, together with the Copernican

Principle implies the Cosmological Principle. The Milky Way is clearly not isotropic, as anyone will know who has looked at the night sky. It occupies a distinct band across the heavens. A Universe consisting only of the Milky Way could therefore not be consistent with the Cosmological Principle.

Although the name 'Cosmological Principle' sounds grand, one should have no illusions about its origin. More often than not, Principles are introduced in order to allow some progress to be made when one has no data to go on. Cosmology was no exception to this rule. It is now known that this guess was basically correct. In the 1920s it was established that the nebulae were definitely outside the Milky Way, and more recent observational studies of the large-scale distribution of galaxies and the cosmic microwave background (discussed in Chapter 7) seem to indicate the Universe is smooth on large scales as required by this idea. It is only more recently that astrophysicists have come up with a reasonably convincing argument as to why the Universe has this special symmetry. The mysterious origin of large-scale smoothness has been called the horizon problem and it is one of the issues addressed by the idea of cosmic inflation discussed in Chapter 8.

Einstein's biggest blunder

Armed with the Cosmological Principle, Einstein was able to construct self-consistent mathematical models of the Universe. Immediately, however, he ran into a problem. It was an unavoidable consequence of his theory that, in any solution of his equations in which the Cosmological Principle applies, space-time must be dynamic. This meant that it was impossible for him to construct a model of a cosmos that is static and unchanging with time. His theory required the Universe to be either expanding or contracting, although it didn't say which of these two possibilities would be the case. Einstein didn't have a great knowledge of astronomy, but he had asked experts about the motions of distant stars. Perhaps because he asked the wrong question, he got the answer that on average the stars were neither approaching nor receding from the Sun. This is actually true inside our galaxy, but we now know it is not the case for other galaxies.

Einstein was so convinced that the Universe should be static that he went back to his original equations. He realized that he could retain their essential character but introduce a slight modification that would counteract the tendency for his cosmological models to expand or contract with time. The modification he introduced was called the 'cosmological constant'. This new term in the theory represents an alteration of the behaviour of gravity on the very largest scales. The cosmological constant allows space itself to possess a tendency to expand or contract, and it can be adjusted in the theory so that it exactly balances the expansion or contraction the Universe would otherwise be forced to possess.

Satisfied with this fix for the time being, Einstein went on to construct a static cosmological model, which he published in 1917. Some years later, in 1929, Hubble published the results that led to the acceptance of the idea that the Universe was not static after all, but expanding. Einstein's original model is now only of historical interest. Without the need to prevent global expansion, there was no need for him to have introduced the cosmological constant. In his later years, Einstein referred to this episode as the greatest blunder he had made in science. This comment is usually taken to refer to the cosmological constant itself, but the true blunder was to have failed to predict the expansion of the Universe.

Although until recently most cosmologists were happy not to include it in their models, the cosmological constant never really went away. It lurked in the background like a mad relative living in the attic. Now, as we shall see in later chapters, it has broken free from obscurity and again plays a leading role. For the remainder of this chapter, however, I shall put it to one side.

The Friedmann models

Einstein was not the only scientist to turn to cosmology in the years immediately following the publication of the general theory of relativity in 1915. One of the others was an obscure Russian physicist by the name of Alexander Friedmann. It was not Einstein but Friedmann who developed the mathematical models of an expanding Universe, which form the basis of the modern Big Bang cosmology. His achievements in this respect are all the more remarkable because he performed his calculations in conditions of extreme hardship during the siege of Petrograd. Friedmann died in 1925, before his work (published in 1922) had achieved any international recognition. Stalin later liquidated the institute he had worked in. Somewhat later a Belgian priest, Georges Lemaître, independently obtained the same results, and it is through Lemaître that these ideas were explored and amplified in Western Europe.

The simplest Friedmann models are the special family of solutions to Einstein's equations obtained by requiring that the Cosmological Principle holds, and assuming that there is no cosmological constant. The Cosmological Principle plays a big part in these models. In relativity theory, time and space are not absolutes. The mathematical description of these two aspects of events (the 'when' and the 'where') involves a complicated four-dimensional 'space-time', which is hard to conceptualize. In general, Einstein's theory does not give an unambiguous way of separating space and time. Different observers can disagree about the time elapsed between events, depending on their motion and on the gravitational fields they have experienced. If the Cosmological Principle applies then there is a special way to think about time that makes all this much simpler. If the Universe has the same density everywhere (which it must if it is homogeneous) then the density of matter itself defines a kind of clock. If the Universe expands, then the space between particles increases and the density of matter consequently goes down. The later the time, the lower the density of matter. Likewise, a higher density implies an earlier time. Observers anywhere in the Universe can set their clocks according to the local density of matter so that all these clocks will be perfectly synchronized. and a perfect synchronization can therefore be achieved. The measure of time that results is usually called 'cosmological proper time'.

Because the density is the same in every place, and it is the density of matter and/or energy that determines the curvature of space through the Einstein equations, the Cosmological Principle also simplifies the way space can curve in response to gravity. Space can be warped, but it must be warped in the same way at every point. There are in fact only three ways in which this can happen.

The obvious way of having the same curvature at every point is to have no curvature at every point. This is usually called the *flat* universe. In a flat universe, light travels in straight lines and all the laws of Euclidean geometry apply just as they did in the 'normal' world. But if space isn't curved, what has happened to gravity? There is matter in a flat universe so why does it not warp space? The answer is that the mass of the universe does warp space, but this is exactly counterbalanced by energy contained in the expansion of the Universe; matter and energy conspire to negate each other's gravitational effects. And in any case, even though space may be flat, space–time is still *curved*.

The flat universe is clearly special because it requires an exact balance between the expansion and the gravitational pull of the matter. When these are not in balance, there are two other alternatives. If the universe has a high matter density then the gravitational effect of the mass within it wins, and it pulls space in on itself like a three-dimensional version of the surface of a sphere. Mathematically, the curvature of space is positive in such a situation. In this case, the *closed universe*, light rays actually converge on each other. While a flat universe can extend indefinitely in all directions, the closed universe is finite. Go off in one direction and you will come back to where you started from. The other alternative is the *open universe*. This too is infinite, but is harder to visualize than the closed version because the space curvature is negative. Light rays diverge in this example, as illustrated in the two-dimensional example shown in the Figure 5.

The behaviour of space in these models mirrors the way they evolve in time. A closed universe is a finite space, but it also has a finite duration. If the universe is expanding at any time and is closed, the expansion will slow down in the future. Eventually the universe will stop expanding and recollapse. The open and flat models will expand for ever. Gravity always fights the expansion of the universe in the Friedmann models, but only in the closed model does it actually win.

The Friedmann models underpin much of the modern Big Bang theory, but they also contain the key to its greatest weakness. If we use these calculations to reverse the expansion of the Universe and turn the clock back on the current state of the cosmos, we find the universe gets ever denser the earlier we go. If we try to go back too far, the mathematics fall apart at a *singularity*.

The singular nature of gravity

In mathematics, a singularity is a pathological property wherein the numerical value of a particular quantity becomes infinite during the course of a calculation. To give a very simplified example, consider the calculation of the Newtonian force due to gravity exerted by a massive body on another particle. This force is inversely proportional to the square of the distance between the two bodies, so that if one tried to calculate the force for objects at zero separation, the result would be infinite. Singularities are not always signs of serious mathematical problems. Sometimes they are simply caused by an inappropriate choice of coordinates. For example, something strange and akin to a singularity happens in the standard maps to be found in an atlas. These maps look quite sensible until you look very near the poles. In a

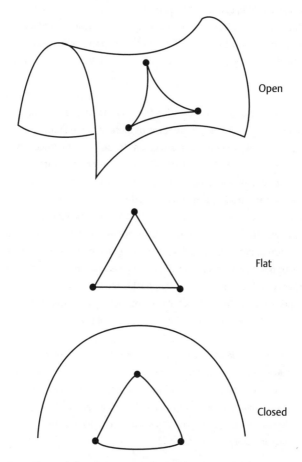

5. Open, flat, and closed spaces in two dimensions. In a flat two-dimensional space (middle) the laws of Euclidean geometry hold true. In this case the sum of the angles of a triangle is 180 degrees. In a closed space such as a sphere (bottom), the angles of a triangle come to somewhat more than 180 degrees, whereas for an open space (such as the saddle shape shown) it is less than 180 degrees.

standard equatorial projection, the North Pole does not appear as a point as it should, but is spread out from a point to a straight line along the top of the map. But if you were to travel to the North Pole you would not see anything catastrophic there. The singularity that causes this point to appear is an example of a coordinate singularity, and it can be transformed away by using a different kind of projection. Nothing particularly odd will happen to you if you attempt to cross this kind of singularity.

Singularities occur with depressing frequency in solutions of the equations of general relativity. Some of these are coordinate singularities like the one I just discussed. These are not particularly serious. However, Einstein's theory is special in that it predicts the existence of real singularities where real physical quantities that should know better, such as the density of matter or the temperature, become infinite. The curvature of space-time can also become infinite in certain situations. The existence of these singularities suggests to many that some fundamental physics describing the gravitational effect of matter at extreme density is absent from our understanding. It is possible that a theory of quantum gravity might enable physicists to calculate what happens deep inside a black hole without having all mathematical quantities becoming infinite. Indeed, Einstein himself wrote in 1950:

> The theory is based on a separation of the concepts of the gravitational field and matter. While this may be a valid approximation for weak fields, it may presumably be quite inadequate for very high densities of matter. One may not therefore assume the validity of the equations for very high densities and it is just possible that in a unified theory there would be no such singularity.

Probably the most famous example of a singularity lies at the centre of a black hole. This appears in the original Schwarzschild solution corresponding to a hole with perfect spherical symmetry. For many years, physicists thought that the existence of a singularity of this kind was merely due to the rather artificial special nature of this spherical solution. However, in a series of mathematical investigations, Roger Penrose and others have shown that no special symmetry is required and that singularities arise whenever any objects collapse under their own gravity.

As if to apologize for predicting these singularities in the first place, general relativity does its best to hide them from us. A Schwarzschild black hole is surrounded by an event horizon that effectively protects outside observers from the singularity itself. It seems likely that all singularities in general relativity are protected in this way, and so-called *naked singularities* are not thought to be physically realistic.

In the 1960s, however, Roger Penrose's work on mathematical properties of the black hole singularity came to the attention of Stephen Hawking, who had the idea of trying to apply them elsewhere. Penrose had considered what would happen in the future when an object collapses under its own gravity. Hawking was interested to know whether these ideas could be applied instead to the problem of understanding what had happened in the past to a system now known to be expanding, i.e. the Universe! Hawking contacted Roger Penrose about this, and they worked together on the problem of the cosmological singularity, as it is now known. Together they showed that expanding-universe models predict the existence of a singularity at the very beginning, where the temperature and density become infinite. No matter whether the universe is open, closed, or flat, there is a fundamental barrier to our understanding. In the beginning, there was infinity.

Most cosmologists interpret the Big Bang singularity in much the same way as the black hole singularity discussed above, i.e. as meaning that Einstein's equations break down at some point in the early Universe due to the extreme physical conditions present there. If this is the case, then the only hope for understanding the early stages of the expansion of the

Universe is through a better theory. Since we don't have such a theory, the Big Bang is incomplete. In particular, since we need to know the total energy budget of the Universe to know whether it is open or closed, we cannot determine by theory alone which of these alternatives is the 'correct' description. This shortcoming is the reason why the word 'model' is probably more appropriate than 'theory' for the Big Bang. The problem of not knowing about the initial conditions of the Universe is the reason why cosmologists still cannot answer some basic questions, such as whether the Universe will expand forever.

Chapter 4
The expanding Universe

So far I have concentrated on the way in which developments in
theoretical physics, particularly the general theory of relativity, led to
major developments in cosmological theory in the 1920s. But these
new ideas only gained acceptance when improved observational
facilities allowed astronomers to begin making reliable estimates of
the distances to and motions of galaxies. In this chapter I will discuss
these observations, and how they fit into the theoretical
framework.

Hubble's Law

The nature of the expansion of the Universe is encapsulated in one
simple equation, now known as the Hubble Law. This states that the
apparent velocity v of a galaxy away from the observer is proportional
to its distance d. Nowadays the constant of proportionality is known
as the Hubble constant and is given the symbol H or H_o. The Hubble Law
is thus written '$v = H_o d$'. The relationship between v and d is called a
linear relationship because if you plot a graph (like Hubble did) of the
measured velocities and distances of a sample of galaxies, you find
they lie on a straight line. The slope of this line is H_o. The Hubble Law
basically means that galaxies twice as far away from the observer are
moving away twice as quickly. Those three times away move three
times as fast, and so on.

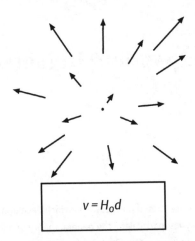

$$v = H_0 d$$

6. Hubble's Law. As observed from the central point, Hubble's Law states that the apparent recession velocity of distant galaxies is proportional to their distance, so the further away they are the quicker they recede. The expansion does not have a centre: any point can be treated as the origin.

Hubble published the discovery of his famous law in 1929, which resulted from a study of the spectra of a sample of galaxies. The American astronomer Vesto Slipher also deserves a large part of the credit for the discovery. As early as 1914 Slipher had obtained spectra of a group of nebulae (as galaxies were then called) that also displayed this relationship, although his distance estimates were very rough. Unfortunately, Slipher's early results, presented at the 17th Meeting of the American Astronomical Association in 1914, were never published; history has never adequately acknowledged the contribution Slipher made.

So how did Hubble obtain his law? The technique he used is called spectroscopy. Light from a galaxy contains a mixture of colours, produced by all the stars within it. A spectroscope splits light up into its component hues so that its precise mixture of colours can be analysed separately. A prism is a simple way of achieving the same end. With a

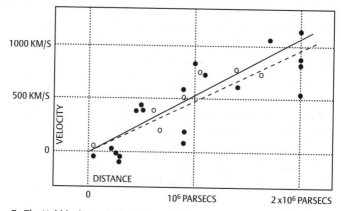

7. The Hubble diagram. Hubble's original velocity–distance plot published in 1929. Notice that some nearby galaxies are actually approaching the galaxy, and there is considerable scatter in his plot.

prism ordinary white light can be split into a spectrum that resembles a rainbow. But as well as having different colours, astronomical spectra also contain sharp features called emission lines. These lines are produced in the gas contained in an object by electrons shifting between different energy levels. These transitions occur at definite wavelengths depending on the chemistry of the source; these wavelengths can be measured accurately in laboratory experiments. Hubble was able to identify emission lines in many of his galaxies. But comparing their position in the measured spectrum to where the lines should be, he found they were usually in the wrong place. In fact, the lines were almost always shifted to the red end of the spectrum, towards longer wavelengths. Hubble interpreted this as a Doppler shift.

Doppler shift

The Doppler effect was originally introduced to physics with a fanfare in the 1840s. In fact this is literally true, because the first experimental demonstration of this effect involved several trumpeters moving on a steam train. The application in that instance was to the properties of sound waves when there is relative motion between the source of the sound and the receiver. We are all familiar with the effect from everyday experience: an approaching police siren has a higher pitch than a receding one. The easiest way to understand the effect is to remember that the pitch of sound depends on the wavelength of the waves from which it is made. High pitch means short wavelength. If a source is travelling near the speed of sound, it tends to catch up the waves it emits in front, thus reducing their apparent wavelength. Likewise, it tends to rush ahead of the waves it emits behind, increasing the gap between the waves and thus lowering their apparent pitch.

In the astronomical setting, the Doppler effect applies to light. Usually the effect is very small, but it becomes appreciable if the velocity of a source is a significant fraction of the velocity of light. (The Doppler effect for sound is small unless the speed of the car is reasonably large, the relevant scale being set by the speed of sound.) A moving source of emission tends to produce light of shorter wavelength if it is approaching the observer and longer wavelength if it is receding. In these cases the light is shifted towards the blue and red parts of the spectrum, respectively. In other words, there is a blueshift (approaching source) or a redshift (receding source).

If the source is emitting white light, however, one would not be able to see any kind of shift. Suppose each line were redshifted by an amount x in wavelength. Then light emitted at a wavelength y would be observed at wavelength $y + x$. But the same amount of light would still be observed at the original wavelength y, because light originally emitted at wavelength $y - x$ would be shifted there to fill the gap. White light therefore still looks white, regardless of the Doppler shift. To see an effect, one has to look at emission lines, which occur at discrete frequencies so that no such compensation can occur. A whole set of lines will be shifted one way or the other in the spectrum, but the lines will keep their relative spacing and it is therefore quite easy to identify how far they have shifted relative to a source which is at rest in a laboratory.

Hubble measured a larger redshift for the more distant galaxies in his sample than for the nearby ones. He assumed that what he was seeing was a Doppler shift, so he converted the shifting of the spectral lines into a measure of velocity. When he plotted this 'apparent recession velocity' against the distance, he got his famous linear relationship. Although Hubble's Law is now taken to represent the expansion of the Universe, Hubble himself never made this interpretation of his results. Lemaître was probably the first theorist to explain Hubble's Law in terms of the expansion of the entire Universe. Lemaître's paper, published in 1927, prefiguring Hubble's classic paper of 1929, had made little impression at the time because it was written in French and published in an obscure Belgian journal. It was not until 1931 that the British astronomer Arthur Stanley Eddington had Lemaître's paper published (in English) in the more influential *Monthly Notices of the Royal Astronomical Society*. The identification of the Hubble Law with the cosmic expansion is one of the main supporting pillars of the Big Bang theory, so Lemaître too deserves great credit for making this important step.

Interpreting the Hubble Law

The fact that galaxies are observed to be moving away from us suggests that we must be at the centre of the expansion. Doesn't this violate the Copernican Principle and put us in a special place? The answer is 'no'. Any other observer would also see everything moving away. In fact, every point in the Universe is equivalent as far as the expansion is concerned. Moreover it can be proved mathematically that Hubble's Law *must* apply in a homogeneous and isotropic expanding universe, i.e. one in which the Cosmological Principle holds. It is the only way such a universe can expand.

It may help to visualize the situation by reducing the three dimensions of space to the two-dimensional surface of a balloon (this would be a closed universe, but the geometry does not particularly matter for this illustration). If one paints dots onto the surface of the balloon and then blows it up, each dot sees all the other dots moving away as if it were the centre of expansion. This analogy has a problem, however, in that one tends to be aware that the two-dimensional surface is embedded in the three dimensions of our ordinary space. One therefore sees the centre of the space inside the balloon as the real centre of expansion. This is inaccurate. One must think of the balloon as being the entire Universe. It is not embedded in another space and there is no such global centre. Every point in the balloon is the centre. This difficulty is often also confused in one's mind with the question of where the Big Bang actually happened: are we not moving away from the site of the original explosion? Where was this explosion situated? The answer to this is the explosion happened everywhere and everything is moving away from it. But in the beginning, at the Big Bang singularity, everywhere and everything was in the same place.

More than seventy years after Lemaître, the Hubble Law still poses some difficulties of interpretation. Hubble had not measured velocities but redshifts. The redshift, usually given the symbol z in cosmology, measures the fractional change in wavelength of an observed line relative to its expected position. Hubble's Law is sometimes stated as a linear relationship between redshift z and distance d, rather than between recession velocity v and d. If the velocities concerned are much smaller than the speed of light c then there is no problem because in this case the redshift is roughly the velocity of the sources expressed as a fraction of the speed of light. So if z and d are proportional and so are z and v, then v and d are also. But when the redshifts are large this relationship breaks down. What, then, is the correct form to use? In the Friedmann models, the interpretation of Hubble's Law is amazingly simple. The linear relationship between recession velocity v and distance d, is *exact* even when the velocity is arbitrarily large.

This may worry some of you, because you will have heard that it is not possible for objects to move faster than light. In a Friedmann universe the more distant is the object, the greater its velocity away from the observer. The velocity of the object can exceed the speed of light by any amount you please. It does not violate any principle of relativity, however, because the observer cannot see it; it is infinitely redshifted.

There is also a potential problem in what is meant by d and how to measure it. Astronomers cannot usually measure the distance of an object directly. They cannot extend a ruler to a distant galaxy and cannot usually use triangulation like surveyors do because the distances involved are too large. They have instead to make measurements using light emitted by the object. Since light travels with a finite speed and, as we know thanks to Hubble, the Universe is expanding, things are not at the same position now as they were when light set out from them. Astronomers are therefore forced to use indirect distance measurements, and to attempt to correct for the expansion of the Universe to locate where the object actually is.

But in fact the theory helps here too. Thinking about velocities and distances of sources is unnecessarily complicated. While the redshift is usually thought of as a Doppler shift, there is another way of picturing this effect, which is much simpler and actually more accurate. In the expanding universe, separations between any points increase uniformly in all directions. Imagine an expanding sheet of graph paper. The regular grid on the paper at some particular time will look like a blown-up version of the way it looked at an earlier time. Because the symmetry of the situation is preserved, one only needs to know the factor by which the grid has been expanded in order to recover the past grid from the later one. Likewise, since a homogeneous and isotropic universe remains so as it expands, one only needs to know an overall 'scale factor' to obtain a picture of the past physical conditions from present data. This factor is usually given the symbol $a(t)$ and its behaviour is governed by the Friedmann equations discussed in the previous chapter.

Remember that light travels with a finite speed. Light arriving now from a distant source must have set out at some finite time in the past. At the time of emission the Universe was younger than it is now and, since it has been expanding, it was smaller then too. If the Universe has expanded by some factor between the emission of light and its detection at a telescope, the light waves emitted would be stretched by the same factor as they travelled through space. For example, if the universe expanded by a factor three then the wavelength would triple. This is a 200 per cent increase and the source consequently is observed to have redshift 2. If the expansion factor were only by 10 per cent (i.e. a factor 1.1) then the redshift would be 0.1, and so on. The redshift is due to the stretching of space-time caused by cosmic expansion.

This interpretation is so simple that it eluded physicists for many years. In 1917, Wilhem de Sitter had published a cosmological model in which he found light rays would be redshifted. Because he had used strange coordinates in which to express his results, he didn't realize his model

Redshift

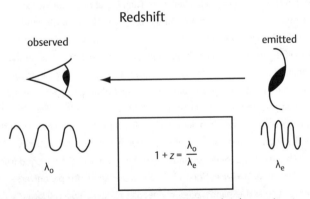

observed

emitted

$$1 + z = \frac{\lambda_o}{\lambda_e}$$

λ_o

λ_e

8. Redshift. As light travels from a source galaxy to the observer it gets stretched by the expansion of the Universe, eventually arriving with a longer wavelength than when it started.

represented an expanding Universe and instead he sought to explain what he had found as some kind of weird gravitational effect. There was considerable confusion about the nature of the 'de Sitter effect' for many years, but it is now known to be extremely simple.

It is important also to stress that not everything takes part in the expansion. Objects that are held together by forces other than gravity do not participate. This includes elementary particles, atoms, molecules, and rocks. These instead remain at a fixed physical size as the Universe swells around them. Likewise, objects in which the force of gravity is dominant also resist the expansion. Planet, stars, and galaxies are bound so strongly by gravitational forces that they are not expanding with the rest of the Universe. On scales even larger than galaxies, not all objects are moving away from each other either. For example, the Andromeda galaxy (M31) is actually approaching the Milky Way because these two objects are held together by their mutual gravitational attraction. Some massive clusters of galaxies are similarly held together against the cosmic flow. Objects larger than this may not necessarily be bound (like individual galaxies are), but their gravity may still be strong enough to cause a distortion of Hubble's Law. Although the linearity of the Hubble Law is now well established out to quite large distances, there is considerable 'scatter' about the straight line. Part of this represents statistical errors and uncertainties in the distance measurements, but this is not the whole story. Hubble's Law is only exactly true for objects moving in an idealized homogeneous and isotropic Universe. Our Universe may be roughly like this on large enough scales, but it is not exactly homogeneous. Its clumpiness deflects galaxies from the pure 'Hubble flow' causing the scatter in Hubble's plot.

But on the largest scales of all, there are no forces strong enough to counteract the global tendency of the Universe to expand with time. In a broad-brush sense, therefore, ignoring all these relatively local perturbations, all matter is rushing apart from all other matter with a velocity described by Hubble's Law.

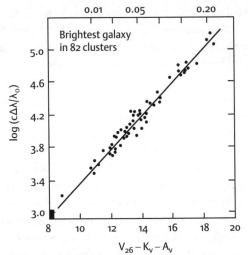

9. The Hubble diagram updated. A more recent compilation of velocities and distances based on work by Allan Sandage. The distance range covered is much greater than in Hubble's original diagram. The small black rectangle in the bottom left of the diagram would entirely cover Hubble's 1929 data.

The quest for H_o

So far I have concentrated on the form of the Hubble Law, and how it is interpreted theoretically. There is one other important aspect of Hubble's Law to discuss, and that is the value of the constant H_o. The Hubble constant H_o is one of the most important numbers in cosmology, but it is also an example of one of the failings of the Big Bang model. The theory cannot predict what value this important number should take; it is part of the information imprinted at the beginning of the Universe where our theory breaks down. Obtaining a true value for H_o using observations is a very complicated task. Astronomers need two measurements. First, spectroscopic observations reveal the galaxy's redshift, indicating its velocity. This part is relatively straightforward. The second measurement, that of the distance, is by far the more difficult to perform.

Suppose you were in a large dark room in which there is a light bulb placed at an unknown distance from you. How could you determine its distance? One way would be to attempt to use some kind of triangulation. You could use a surveying device such as a theodolite, moving around in the room, measuring angles to the bulb from different positions, and using trigonometry to work out the distance. An alternative approach is to measure distances using the properties of the light emitted by the bulb. Suppose you knew that the bulb was, say, a 100-watt bulb. Suppose also that you were equipped with a light meter. By measuring the amount of light you receive using the light meter, and remembering that the intensity of light falls off as the square of the distance, you could infer the distance to the bulb. If you didn't know in advance the power of the bulb, however, this method would not work. On the other hand, if there were two identical bulbs in the room with unknown but identical wattage then you could tell the relative distances between them quite easily. For example, if one bulb produced a reading on your light meter that was four times smaller than the reading produced by the other bulb then the first bulb must be twice as far away as the second. But you still don't know in absolute terms how far it is to either of the bulbs.

Putting these ideas into an astronomical setting highlights the problems of determining the distance scale of the Universe. Triangulation is difficult because it is not feasible to move very much relative to the distances concerned, except in special situations (see below). Measuring absolute distances using stars or other sources is also difficult unless we can find some way of knowing their intrinsic luminosity (or power output). A feeble star nearby looks the same as a very bright star far away, since stars, in general, cannot be resolved even by the most powerful telescopes. If we know that two stars (or other sources) are identical, however, then measuring relative distances is not so difficult. It is the calibration of these relative distance measures that forms the central task of work on the extragalactic distance scale.

To put these difficulties into perspective, one should remember that it was not until the 1920s that there was even a rough understanding of the scale of the Universe. Prior to Hubble's discovery that the spiral nebulae (as they were then called) were outside the Milky Way, the consensus was that the Universe was actually very small indeed. These nebulae, now known to be spiral galaxies like the Milky Way, were usually thought to represent the early stages of formation of structures like our Solar System. When Hubble announced the discovery of his eponymous law, the value of H_0 he obtained was about 500 kilometres per second per Megaparsec (the usual units in which the Hubble constant is measured). This is about eight times larger than current estimates. Hubble had made a mistake in identifying a kind of star to use as a distance indicator (see below) and, when his error was corrected in the 1950s by Baade, the value dropped to about 250 in the same units. Sandage, in 1958, revised the value still further to between 50 and 100 and present observational estimates still lie in this range.

Modern measurements of H_0 use a battery of distance indicators, each one taking one step upwards in scale, starting with local estimates of distances to stars within the Milky Way, and ending at the most distant galaxies and clusters of galaxies. The basic idea, however, is still the same as that pioneered by Hubble and Sandage.

First, one exploits local kinematic distance measures to establish the scale of the Milky Way. Kinematic methods do not rely upon knowledge of the absolute luminosity of a source, and they are analogous to the idea of triangulation mentioned above. To start with, distances to relatively nearby stars can be gauged using the *trigonometric parallax* of a star, i.e. the change in the star's position on the sky in the course of a year due to the Earth's motion in space. The usual astronomers' unit of distance – the parsec (pc) – stems from this method: a star one parsec away produces a parallax of one second of arc when the Earth moves from one side of the Sun to the other. For reference, one parsec is around three light years. The important astrometric satellite

Hipparchos was able to obtain parallax measurements for thousands of stars in our galaxy.

Another important class of distance indicators contains variable stars of which the most important are the Cepheid variables. The variability of these objects gives clues about their intrinsic luminosity. The classical Cepheids are bright variable stars known to display a very tight relationship between the period of variation P and their absolute luminosity L. The measurement of P for a distant Cepheid thus allows one to estimate its L, and hence its distance. These stars are so bright that they can be seen in galaxies outside our own and they extend the distance scale to around 4 Mpc (4,000,000 pc). Errors in the Cepheid distance scale, due to interstellar absorption, galactic rotation, and, above all, a confusion between Cepheids and another type of variable star, called the W Virginis variables, were responsible for Hubble's large original value for H_o. Other stellar distance indicators allow the ladder to be extended slightly to around 10 Mpc. Collectively, these methods are given the name *primary distance indicators*.

The *secondary distance indicators* include HII regions (large clouds of ionized hydrogen surrounding very hot stars) and globular clusters (clusters of around one hundred thousand to ten million stars). The former of these has a diameter, and the latter an absolute luminosity, which has a small scatter around the mean for these objects. With such relative indicators, calibrated using the primary methods, one can extend the distance ladder out to about 100 Mpc. The *tertiary distance indicators* include brightest cluster galaxies and supernovae. Clusters of galaxies can contain up to about a thousand galaxies. One finds that the brightest elliptical galaxy in a rich cluster has a very standard total luminosity, probably because these objects are known to be formed in a special way by cannibalizing other galaxies. With the brightest galaxies one can reach distances of several hundred Mpc. Supernovae are stars that explode, producing a luminosity roughly equal to that of an entire galaxy. These stars are therefore easily seen in distant galaxies. Many other indirect distance estimates have also been explored, such as correlations between various intrinsic properties of galaxies.

So there seems to be no shortage of techniques for measuring H_o. Why is it then that the value of H_o is still known so poorly? One problem is that a small error in one rung of the distance ladder also affects higher levels of the ladder in a cumulative way. At each level there are also many corrections to be made: the effect of galactic rotation in the Milky Way; telescope aperture variations; absorption and obscuration in the Milky Way; and observational biases of various kinds. Given the large number of uncertain corrections, it is perhaps not surprising that we are not yet in a position to determine H_o with any great precision. Controversy has surrounded the distance scale ever since Hubble's day. An end to this controversy seems to be in sight, however, because of the

10. The Hubble Space Telescope. This photograph was taken as the shuttle was deployed from the Space Shuttle in 1990. One of the most important projects the Hubble Telescope has undertaken has been to measure distances to stars in distant galaxies in order to measure Hubble's constant.

latest developments in technology. In particular, the Hubble Space Telescope (HST) is able to image stars, particularly Cepheid variables, directly in galaxies within the Virgo cluster of galaxies, an ability which bypasses the main sources of uncertainty in the calibration of traditional steps in the distance ladder. The HST key programme on the distance scale is expected to fix the value of Hubble's constant to an accuracy of about 10 per cent. This programme is not yet complete, but latest estimates are settling on a value of H_o in the range 60 to 70 kilometres per second per Megaparsec.

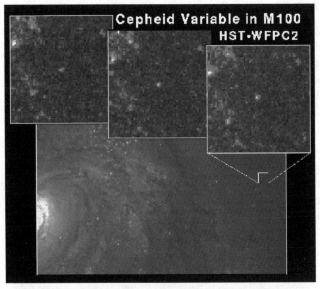

Cepheid Variable in M100
HST·WFPC2

11. Cepheids in M100. These pictures were taken with the Hubble Telescope; the three images indicate the presence of a variable star now known to be a Cepheid. Hubble has been able to measure the distance to this galaxy, directly bypassing the indirect methods in use prior to the launch of this telescope.

The age of the Universe

If the expansion of the Universe proceeded at a constant rate then it would be a very simple matter to relate the Hubble constant to the age of the Universe. All the galaxies are now rushing apart, but in the beginning they must have all been in the same place. All we need to do is work out when this happened; the age of the Universe is then the time elapsed since this event. It's an easy calculation, and it tells us that the age of the Universe is just the inverse of the Hubble constant. For current estimates of H_o the age of the Universe works out around 15 billion years.

This calculation would only be true, however, in a completely empty universe that contained no matter to cause the expansion to slow down. In the Friedmann models, the expansion is decelerated by an amount depending on how much matter there is in the Universe. We don't really know exactly how much deceleration needs to be allowed for, but it's clear the age will always be less than the value we just calculated. If the expansion is slowing down, it must have been faster in the past, so the universe must have taken less time to get to where it is. The effect of deceleration is, however, not particularly large. The age of a flat universe should be about 10 billion years.

An independent method for estimating the age of the Universe is to date objects within it. Obviously, since the Big Bang represents the origin of all matter as well as of space-time, there should be nothing *in* the Universe that is older *than* the Universe. Dating astronomical objects is, however, not easy. One can estimate ages of terrestrial rocks using the radioactive decay of long-lived isotopes, such as uranium-235, which have a half-life measured in billions of years. The method is well understood and similar to the archaeological use of radiocarbon dating, with the only difference being the vastly larger timescale needed for the cosmological application requiring the use of elements with much longer half-lives than carbon-14. The limitation of such approaches,

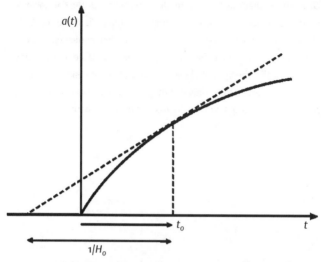

12. The age of the Universe. Whether they are open, flat, or closed the usual Friedmann models are always slowing down. This means that the Hubble time, $1/H_o$, always exceeds the actual time elapsed since the Big Bang (t_o).

however, is that they can only be used to date material within the Solar System. Lunar and meteoritic rocks are older than terrestrial material, but they may have formed very recently indeed during the history of the Universe so are not useful in the cosmological setting

The most useful method for measuring the age of the Universe is less direct. The strongest constraints come from studies of globular star clusters. The stars in these clusters are thought to have all formed at the same time, and the fact that they generally are stars of very low mass suggests that they are quite old. Because they all formed at the same time, a collection of these stars can be used to calculate how they have been evolving. This puts a lower limit on the age of the Universe, because one must allow some time since the Big Bang to form the clusters in the first place. Recent studies suggest that such systems are around 14 billion years old, though this has become controversial in recent years.

One can see that this poses immediate problems for the flat universe model. Globular cluster stars are simply too old to fit in the short lifetime of such a universe. This argument has lent some support to the argument that we in fact live in an open universe. More recently, and more radically, the ages of old stars also seem to fit neatly with other evidence suggesting the Universe might have been been speeding up rather than slowing down. I'll discuss this more in Chapter 6.

Chapter 5
The Big Bang

While the basic theoretical framework of the Friedmann models has been around for many years, the Big Bang has emerged only relatively recently as the likeliest broad-brush account of how the contents of the Universe have evolved with time. For many years, most cosmologists favoured an alternative model called the Steady State model. Indeed, the Big Bang itself had a number of variants. A more precise phrasing of the modern theory is to call it the 'hot Big Bang' to distinguish it from an older rival (now discarded), which had a cold initial phase. As I have mentioned already, it is also not entirely correct to call this a 'theory'. The difference between theory and model is subtle, but a useful definition is that a theory is usually expected to be completely self-contained (it can have no adjustable parameters, and all mathematical quantities are defined a priori) whereas a model is not complete in the same way. Owing to the uncertain initial stages of the Big Bang, it is difficult to make cast-iron predictions and it is consequently not easy to test. Advocates of the Steady State theory have made this criticism on many occasions. Ironically, the term 'Big Bang' was initially intended to be derogatory and was coined in a BBC radio programme by Sir Fred Hoyle, one the model's most prominent dissidents.

Steady State theory

In the Steady State cosmological model, advanced by Gold, Hoyle, Bondi, and Narlikar (amongst others), the universe is expanding but nevertheless has the same properties at all times. The principle behind this theory is called the Perfect Cosmological Principle, a generalization of the Cosmological Principle that says that the Universe is homogeneous and isotropic in space to include also homogeneity with respect to time.

Because all the properties of Steady State cosmology have to be constant in time the expansion rate of this model is also a constant. It is possible to find a solution of the Einstein equations that corresponds to this. It is called the de Sitter solution. But if the Universe is expanding, the density of matter must decrease with time. Or must it? The Steady State theory postulates the existence of a field, called the C-field, which creates matter at a steady rate to counteract the dilution caused by cosmic expansion. This process, called *continuous creation*, has never been observed in the laboratory but the rate of creation required is so small (about one atom of hydrogen per cubic metre in the age of the Universe) that it is difficult to rule out continuous creation as a possible physical process by direct observation.

The Steady State was a better theory in the eyes of many theorists because it was easier to test than its rivals. In particular, any evidence at all that the Universe was different in the past to what it is like now would rule out the model. From the late 1940s onward, observers attempted to see if the properties of distant galaxies (which one sees as they were in the past) were different from nearby ones. Such observations were difficult and problems of interpretation led to acrimonious disputes between advocates of the Steady State theory and its rival, exemplified by a bitter feud between the radio astronomer Martin Ryle and Fred Hoyle when the former claimed to have found significant evolution in radio source properties. It was not until the mid-1960s that an

Cosmology

accidental discovery shed independent and crucial light on the argument.

The smoking gun

In the early 1960s two physicists, Arno Penzias and Robert Wilson, were using a curious horn-shaped microwave antenna left over from telecommunications satellite tests to study the emission produced by the Earth's atmosphere. The telescope had been designed to study possible sources of interference that might cause problems for planned satellite communication systems. Penzias and Wilson were surprised to find a uniform background of noise, which would not go away. Eventually, after much checking and the removal of pigeons that had been nesting in their telescope, they accepted that the noise was not going to go away. Coincidentally, just down the road in Princeton New Jersey, a group of astrophysicists including Dicke and Peebles had been trying to get together an experiment to detect radiation produced by the Big Bang. They realized they had been beaten to it. Penzias and Wilson published their result in the *Astrophysical Journal* in 1965, alongside a paper from the Dicke group explaining what it meant. Penzias and Wilson were awarded the Nobel Prize in 1978.

Since its discovery the microwave background has been subject to intense scrutiny, and we now know much more about it than was the case in 1965. Penzias and Wilson had noticed that their noise did not depend on the time of day, which one would expect if it were an atmospheric phenomenon. Indeed the very high degree of uniformity of the microwave background radiation shows that it is not even associated with sources within our galaxy (which would not be distributed evenly on the sky). It is definitely an extragalactic background. More importantly, it is now known that this radiation has a very particular kind of spectrum called a black body. Black-body spectra arise whenever the source is both a perfect absorber and a perfect emitter of radiation. The radiation produced by a black body is often called thermal radiation, because the perfect absorption and emission brings the source and the radiation into thermal equilibrium.

The characteristic black-body spectrum of this radiation demonstrates beyond all reasonable doubt that it was produced in conditions of thermal equilibrium in the very early stages of the primordial fireball. The microwave background is now very cold: its temperature is just less than three degrees above absolute zero. But this radiation has gradually been cooling as an effect of the expansion of the Universe, as each constituent photon suffers a redshift. Turning the clock back to earlier stages of the Universe's evolution, these photons get hotter and carry more energy. Eventually one reaches a stage where the radiation starts

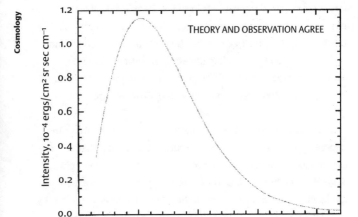

13. The spectrum of the cosmic microwave background. This graph shows the measured intensity of the cosmic microwave background as a function of wavelength. Both theory and measurement are plotted here; the agreement is so good that the two curves lie on top of one another. This perfect black-body behaviour is the strongest evidence that the Universe began with a hot Big Bang.

to have had a drastic effect on matter. Ordinary gas is made from atoms that consist of electrons orbiting around nuclei. In an intense radiation field, however, the electrons are stripped off to form a plasma in which the matter is said to be ionized. This would have happened about 300,000 years after the Big Bang, when the temperature was several thousand degrees and the Universe was about one thousand times smaller and a billion times denser than it is today. At this period the entire Universe was as hot as the surface of the Sun (which, incidentally, also produces radiation of near black-body form). Under conditions of complete ionization, matter (especially the free electrons) and radiation undergo rapid collisions that maintain thermal equilibrium. The Universe is therefore opaque to light when it is ionized. As it expands and cools, the electrons and nuclei recombine into atoms. When this happens, photon scattering is much less efficient. In fact the Universe becomes virtually transparent after recombination, so what we see as the microwave background today is the cool relic radiation that was last scattered by electrons at the epoch of recombination. When it was finally released from scattering processes, this radiation would have been in the optical or ultraviolet part of the spectrum, but since that time it has been progressively redshifted by the expansion of the Universe and is now seen at infrared and microwave wavelengths.

Because of its near-perfect isotropy on the sky, the cosmic microwave background provides some evidence in favour of the Cosmological Principle. It also provides clues to the origin of galaxies and clusters of galaxies. But its importance in the framework of the Big Bang theory far exceeds these. The existence of the microwave background allows cosmologists to deduce the conditions present in the early stages of the Big Bang and, in particular, helps account for the chemistry of the Universe.

Nucleosynthesis

The chemical composition of the Universe is basically very simple. The bulk of known cosmic material is in the form of hydrogen, the simplest of all chemical materials, containing a nucleus of a single proton. More than 75 per cent of the matter in the universe is in this simple form. Aside from the hydrogen, about 25 per cent of the material constituents (by mass) of the Universe is in the form of helium-4, a stable isotope of helium which has two protons and two neutrons in its nucleus. About one hundred thousand times rarer than this come two more exotic elements. *Deuterium*, or heavy hydrogen as it is sometimes called, has a nucleus consisting of one proton and one neutron. The lighter isotope of helium, helium-3, is short of one neutron compared to its heavier version. And finally there comes lithium-7, produced as a tiny trace element with an abundance of one part in ten billion of the abundance of hydrogen. How did this chemical mix come about?

It has been known since the 1930s that stars work by burning hydrogen as a kind of nuclear fuel. As part of this process, stars synthesize helium and other elements. But we know that stars alone cannot be responsible for producing the cocktail of light elements I have just described. For one thing, stellar processes generally involve a destruction of deuterium more quickly than it is produced, because the strong radiation fields in stars break up deuterium into its component protons and neutrons. Elements heavier than helium-4 are made rather easily in stellar interiors but the percentage of helium-4 observed is too high to be explained by the usual predictions of stellar evolution.

It is interesting that the difficulty of explaining the abundance of helium by stellar processes alone was recognized as early as the 1940s by Alpher, Bethe, and Gamow who themselves proposed a model in which nucleosynthesis occurred in the early stages of cosmological evolution. Difficulties with this model, in particular an excessive production of helium, persuaded Alpher and Herman in 1948 to consider the idea that there might have been a significant radiation background at the epoch of nucleosynthesis. They estimated that this background should have a present temperature of around 5 K, not far from the value it is now known to have, although some fifteen years were to intervene before the cosmic microwave background radiation was discovered.

The calculation of the relative amounts of light nuclei produced in the primordial fireball requires a few assumptions to be made about some properties of the Universe at the relevant stage of its evolution. In addition to the normal assumptions going into the Friedmann models, it is necessary also to require that the early Universe went through a stage of thermal equilibrium at temperatures of more than a billion degrees. In the Big Bang model this would have happened very early on indeed, within a few seconds of the beginning. Other than that, the calculations are fairly straightforward and they can be performed using computer codes originally developed for modelling thermonuclear explosions.

Before nucleosynthesis begins, protons and neutrons are continually interconverting by means of weak nuclear interactions (the nuclear interactions are described in more detail a bit later on). The relative numbers of protons and neutrons can be calculated as long as they are in thermal equilibrium and, while the weak interactions are fast enough to maintain equilibrium, the neutron–proton ratio continually adjusts itself to the cooling surroundings. At some critical point, however, the weak nuclear reactions become inefficient and the ratio can no longer adjust. What happens then is that the neutron–proton ratio is 'frozen out' at a particular value (about one neutron for every six protons). This ratio is fundamental in determining the eventual abundance of helium-4. To make helium by adding protons and neutrons together, we first have to make deuterium. But I have already mentioned that deuterium is easily disrupted by radiation. If a deuterium nucleus gets hit by a photon, it falls apart into a proton and neutron. When the Universe is very hot, any deuterium is destroyed as soon as it is made.

This is called the deuterium bottleneck. While this nuclear traffic jam exists, no helium can be made. Moreover, the neutrons which froze out before this start to decay with a lifetime of around ten minutes. The result of the delay is therefore that slightly fewer neutrons are available for the subsequent cooking of helium.

When the temperature of the radiation bath falls below a billion degrees, the radiation is not strong enough to dissociate deuterium and it lingers long enough for further reactions to occur. Two deuterium nuclei can weld together to make helium-3, with the ejection of a neutron. Helium-3 can capture a deuterium nucleus and make helium-4 and eject a proton. These two reactions happen very quickly with the result that virtually all neutrons end up in helium-4, and only traces of the intermediate deuterium and helium-3 are produced. The abundance by mass of helium-4 that comes out naturally is about 25 per cent, just as required. Likewise, the amounts of intermediate nuclei are also close to the observations. All this is done in the first few minutes of the primordial fireball.

This seems like a spectacular success of the theory, which it is indeed. But agreement between detailed calculations of the nuclear fallout from the Big Bang and observed element abundances is only reached for a particular value of one crucial parameter, the baryon-to-photon ratio of the Universe. The whole thing only works if this number is around one in ten billion. That is one proton or neutron for every ten billion photons. We can use the known temperature of the microwave background to work out how many photons there are in the Universe. This can be done very accurately. Since we know the baryon-to-photon ratio required to make nucleosynthesis work we can use the appropriate value to calculate the number of baryons. The result is tiny. The amount of matter in the form of baryons can only be a few per cent of the amount of mass required to close the Universe.

Turning back the clock

The production of the microwave background during the epoch of recombination and the synthesis of the elements during the nuclear fireball are two major successes of the Big Bang theory. The way observations tally with detailed calculations provides firm support for the model. Buoyed by these successes, cosmologists have since tried to use the Big Bang to explore other consequences of matter at very high density and temperature. In this activity, the Big Bang exploits the connection between the world of the very large and that of the very small.

The further into the past we travel, the smaller and hotter the Universe becomes. We are living now at an epoch about 15 billion years after the Big Bang. The microwave background was produced about 300,000 years after the Big Bang. The nuclear furnace did its cooking in the first few minutes. Pushing our understanding of the Universe to earlier times requires knowledge of how matter behaves at energies above those achieved in nuclear reactors. Experiments that can probe such phenomenal scales of energy can only be constructed at enormous cost. Particle accelerators such as those at CERN in Geneva can recreate some aspects of the primeval inferno, but our knowledge of how matter behaves under these extreme conditions is still fragmentary and does not extend to much earlier periods than the epoch of nucleosynthesis.

In the early days physicists saw the Big Bang as a place where they could apply their theories. Now, with theories of particle physics still largely untested elsewhere, it has become a testing-ground. To see how this has happened, we have to understand the development of particle physics over the last forty years.

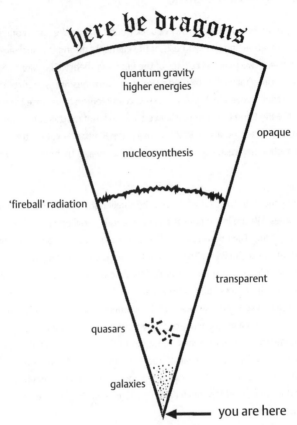

14. Looking back in time. As we look further out into space, we look further back in time. Relatively nearby we see galaxies. Further away we can see highly active galaxies known as quasars. Beyond that there are the 'dark ages': the look-back time is so great that we are viewing the Universe before galaxies formed. Eventually we look so far that the Universe was so hot that it was an opaque fireball much like the central parts of a star. The fireball radiation comes to us through the expanding Universe and arrives as the microwave background. If we could see further than this, we would see nuclear reactions happening, as they do in stars. At earlier times the energies become so high that we have to rely on guesswork. Finally, we reach the edge of the Universe . . . when quantum gravity becomes important we know nothing.

The four forces of nature

Armed with the new theories of relativity and quantum mechanics, and
in many cases further spurred on by new discoveries made possible by
advances in experimental technology, physicists in this century have
sought to expand the scope of science to describe all aspects of the
natural world. All phenomena amenable to this treatment can be
attributable to the actions of the four forces of nature. These four
fundamental interactions are the ways in which the various elementary
particles from which all matter is made interact with each other. I have
already discussed two of these, electromagnetism and gravity. The
other two concern the interactions between the constituents of the
nuclei of atoms, the weak nuclear force and the strong nuclear force.
The four forces vary in strength (gravity is the weakest, and the strong
nuclear force is the strongest) and also in the kinds of elementary
particles that take part in the interactions they control.

The electromagnetic force holds electrons in orbit around atomic nuclei
and is thus responsible for holding together all material with which we
are familiar. However, it was realized early in the twentieth century that,
in order to apply Maxwell's theory in detail to atoms, ideas from
quantum physics and relativity would have to be incorporated. It was
not until the work of Richard Feynman and others, building on the work
of Dirac, that a full quantum theory of the electromagnetic force, called
quantum electrodynamics, was developed. In this theory, usually
abbreviated to QED, electromagnetic radiation in the form of photons is
responsible for carrying the electromagnetic interaction between
particles of different charges.

Before discussing interactions any further it is worth mentioning some
of the properties of the elementary particles between which these
forces act. The basic properties are particles called fermions. These are
distinguished from the force carriers, the bosons (such as the photon),
by their spin. The fermions are divided into two classes, the leptons and

the quarks, each of these classes is divided into three generations, and each generation contains two particles. Altogether, therefore, there are six leptons (arranged in three pairs). One of each lepton pair is charged (the electron is an example), while the other carries no charge and is called a neutrino. While the electron is stable, the other two charged leptons (called the mu and the tau) decay very rapidly and are consequently much more difficult to detect.

The quarks are all charged and the three families of them are also arranged in pairs. The first family contains the 'up' and the 'down'; the second pair is the 'strange' and the 'charmed'; the third contains the 'bottom' and the 'top'. Free quarks are not observed, however. They are always confined into composite particles called hadrons. These particles include the baryons, which are combinations of three quarks, the most familiar examples of which are the proton and the neutron. There are many other hadron states, but most of them are very unstable. They might be produced in accelerator experiments (or in the Big Bang) but do not hang around for long before decaying. Using our current understanding it seems that within one millionth of a second of the beginning of time quarks have sufficient energy to tear themselves free. At earlier times than this, the familiar hadronic particles dissolve into a 'soup' of quarks.

Each of the fermions also has a mirror-image version called its antiparticle. The antiparticle of the electron is the positron; there are also antiquarks and antineutrinos.

The theory of QED describes interactions between the charged fermions. The next force to come under the spotlight was the weak nuclear force, which is responsible for the decay of certain radioactive materials. The weak interaction involves all kinds of fermions including the neutrinos which, being uncharged, cannot feel the QED interaction. As in the case of electromagnetism, weak forces between particles are mediated by other particles – not photons, in this case, but massive

ELEMENTARY PARTICLES

Quarks				Force Carriers
u up	c charm	t top	γ photon	
d down	s strange	b bottom	g gluon	
ν_e electron neutrino	ν_μ mu neutrino	ν_τ tau neutrino	Z Z boson	
e electron	μ muon	τ tauon	W W boson	
I	II	III		

Leptons

Three Generations of Matter

15. **Building blocks of matter.** The standard model of particle physics consists of a relatively small number of basic particles. There are quarks arranged in three generations each of which contains two particles; heavy nuclear particles are made of such quarks. The leptons are arranged in a similar fashion. The quarks and the leptons are fermions, forces between them are mediated by bosons (on the right) called the photon, the gluons, and the weak W and Z bosons.

particles called the W and Z bosons. The fact that these particles have mass (unlike the photon) is the reason why the weak nuclear force has such a short range and its effects are confined to the tiny scales of an atomic nucleus. The W and Z particles otherwise play the same role in this context as the photon does in QED: they, and the photon, are examples of what are known as gauge bosons.

The theory of the strong interactions responsible for holding the quarks together in hadrons is called quantum chromodynamics (or QCD) and it is built upon similar lines to QED. In QCD, there is another set of gauge bosons to mediate the force. These are called gluons; there are eight of them. In addition QCD has a property called 'colour' which plays a similar role to that of electric charge in QED.

The drive for unification

Is it possible, taking a cue from Maxwell's influential unification of electricity and magnetism in the 19th century, to put all QED, the weak interactions, and QCD together in a single overarching theory?

A theory that unifies the electromagnetic force with the weak nuclear force was developed around 1970 by Glashow, Salaam, and Weinberg. Called the electroweak theory, this represents these two distinct forces as being the low-energy manifestations of a single force. When particles have low energy, and are moving slowly, they do feel the different nature of the weak and electromagnetic forces. Physicists say that at high energies there is a symmetry between the electromagnetic and weak interactions: electromagnetism and the weak force appear different to us at low energies because this symmetry is broken. Imagine a pencil standing on its end. When vertical it looks the same from all directions. A random air movement or passing lorry will cause it to topple: it could fall in any direction with equal probability. But when it falls, it falls some *particular* way picking out some specific direction. In the same way, the difference between electromagnetism and weak nuclear forces could be just happenstance, a chance consequence of how the high-energy symmetry was broken in our world.

The electroweak and strong interactions coexist in a combined theory of the fundamental interactions called the *standard model*. It's an amazing success that all the principal particles predicted by the standard model have now been discovered, with only one exception. (A special boson, called the Higgs, is required to explain the masses in the standard model and it has so far defied detection.) This model, however, does not provide a unification of all three interactions in the same way that the electroweak theory does for two of them. Physicists hope eventually to unify all three of the forces discussed so far in a single theory, which would be known as a Grand Unified Theory, or GUT. There are many contenders for such a theory, but it is not known which (if any) is correct.

One idea associated with unified theories is *supersymmetry*. According to this hypothesis, there is an underlying symmetry between the fermions and the bosons, two families which are treated separately in the standard model. In supersymmetric theories, every fermion has a boson 'partner' and vice versa. Quarks have bosonic partners called squarks, neutrinos have sneutrinos and so on. The photon, a boson, has a fermion partner called the photino. The partner of the Higgs boson is the Higgsino, and so on. One of the interesting possibilities of supersymmetry is that at least one of the myriad of particles that one expects to reveal themselves at very high energy might be stable. Could one of these particles make up the dark matter that seems to pervade the Universe?

Baryogenesis

It is clear that the idea of symmetry plays an important role in particle theory. For example, the equations that describe electromagnetic interactions are symmetric when it comes to electrical charge. If one changed all the positive charges into negative charges, and vice versa, Maxwell's equations that describe electromagnetism would still be correct. To put it another way, the choice of assigning negative charge to electrons and positive charges to protons is arbitrary: it could have been done the other way around, and nothing would be different in the theory. This symmetry translates into the existence of a conservation law for charge; electrical charge can be neither created nor destroyed. It seems to make sense that our Universe should not have a net electrical charge: there should be just as much positive charge as negative charge, so the net charge is expected to be zero. This seems to be the case.

The laws of Physics also seem to fail to distinguish between matter and anti-matter. But we know that ordinary matter is much more common than anti-matter. In particular, we know that the number of baryons (protons and neutrons) exceeds the number of anti-baryons. Baryons actually carry an extra kind of 'charge' called their baryon number B. The Universe carries a net baryon number. Like the net electric charge, one would have thought that B should be a conserved quantity. So if B is not zero now, there seems to be no avoiding the conclusion that it can't have been zero at any time in the past. The problem of generating this asymmetry – the problem of baryogenesis – perplexed scientists working on the Big Bang theory for some considerable time.

The Russian physicist Andrei Sakharov in 1967 was the first to work out under what conditions there could actually be a net baryon asymmetry and to show that, in fact, baryon number need not be a conserved quantity. He was able to produce an explanation in which the laws of Physics are indeed baryon-symmetric, and at early times the Universe had no net baryon number, but as it cooled a gradual preference for baryons over anti-baryons emerged. His work was astonishingly prescient, because it was performed long before any unified theories of particle physics were constructed. He was able to suggest a mechanism which could produce a situation in which for every thousand million anti-baryons in the early Universe, there were a thousand million and one baryons. When a baryon and an anti-baryon collide, they annihilate in a puff of electromagnetic radiation. In Sakharov's model, most of the baryons would encounter anti-baryons, and be annihilated in this way. We would eventually be left with a universe containing thousands of millions of photons for every baryon that survives. This is actually the case in our Universe. The cosmic microwave background radiation contains billions of photons for every baryon. The explanation of this is a pleasing example of the interface between particle physics and cosmology, but it is by no means the most dramatic. In the next chapter, I will discuss the idea of cosmic inflation according to which subatomic physics is thought to affect the entire geometry of the Universe.

Chapter 6
What's the matter with the Universe?

Is the Universe finite or infinite? Will the Big Bang end in a Big Crunch? Is space really curved? How much matter is there in the Universe? And what form does this matter take? One would certainly hope that a successful scientific cosmology could provide answers to questions as basic as these. The answers depend crucially upon a number known as Ω (Omega). Astronomers have long grappled with the problem of how to measure Ω using observations of the Universe around us, with only limited success. Dramatic progress in the development and application of new technology now suggests the possibility that the value of Ω may finally be pinned down within the next few years. But there is a sting in the tale. The most recent observations suggest that Ω does not, after all, hold all the answers. The issue of Ω is, however, not entirely an observational one, because the precise value that this quantity takes holds important clues about the very early stages of the Big Bang, and for the structure of our Universe on very large scales. So why is Ω so important and its value so elusive?

The quest for Ω

To understand the role of Ω in cosmology, it is first necessary to remember how Einstein's general theory of relativity relates geometrical properties of space-time (such as its curvature and expansion), to the physical properties of matter (such as its density and state of motion). As I explained in Chapter 3, the application of this complicated theory in cosmology is greatly simplified by the introduction of the Cosmological Principle. In the end, the evolution of the entire Universe is governed by one relatively simple equation, now known as the Friedmann equation.

The Friedmann equation can be thought of as expressing the law of conservation of energy for the Universe as a whole. Energy comes in many different forms throughout nature but only two relatively familiar forms are involved here. A moving object, such as a bullet, carries a type of energy called *kinetic* energy, which depends upon its mass and velocity. Obviously since the Universe is expanding, and all the galaxies are rushing apart, the Universe contains a great deal of kinetic energy. The other form of energy is *potential* energy, which is a little more difficult to understand. Whenever an object is moving and interacting through some kind of force it can gain or lose potential energy. For example, imagine a weight tied on the end of a dangling piece of string. This makes a simple pendulum. If I raise the weight, it gains potential energy because I have to work against gravity to lift it. If I then release the weight the pendulum begins to swing. The weight then picks up kinetic energy and as it drops it loses potential energy. Energy is transferred between the two types in this process, but the total energy of the system is conserved. The weight will swing to the bottom of its arc, where it has no potential energy, but it will still be moving. It will in fact describe a complete cycle, returning eventually to the top of its arc at which point it stops (instantaneously) before starting another swing. At the top, it has no kinetic energy but maximum potential. Wherever the weight is, the energy of this system is constant. This is the law of conservation of energy.

In cosmological terms, the kinetic energy depends crucially on the expansion rate or, in other words, upon the Hubble constant H_o. The potential energy depends on the density of the Universe, i.e. upon how much matter there is per unit volume of the Universe. Unfortunately, this quantity is not known at all accurately: it is even less certain than the value of the Hubble constant. If we knew the mean density of matter and the value of H_o, however, we could calculate the total energy of the Universe. This would have to be constant in time, in accordance with the law of conservation of energy (or, in this context, the Friedmann equation).

Setting aside the technical difficulties that arise when General Relativity is involved, we can now discuss the evolution of the Universe in broad terms using familiar examples from high-school physics. For instance, consider the problem of launching a vehicle from Earth into space. Here the mass responsible for the gravitational potential energy of the vehicle is the Earth. The kinetic energy of the vehicle is determined by the power of the rocket we use. If we give the vehicle only a modest rocket, so that it doesn't move very quickly at launch, then the kinetic energy is small and may be insufficient for the rocket to escape from the attraction of the Earth. Consequently, the vehicle goes up some way and then comes back down again. In terms of energy, what happens is that the rocket uses up its kinetic energy, given expensively at launch, to pay the price in terms of potential energy for its increased height. If we use a bigger rocket, it would go higher before crashing down to the ground. Eventually, we will find a rocket big enough to supply the vehicle with enough energy for it to buy its way completely out of the gravitational field of the Earth. The critical launch velocity here is usually called *escape* velocity: above the escape velocity, the rocket keeps on going for ever; below it the rocket comes crashing down again.

In the cosmological setting the picture is similar but the critical quantity is not the velocity of the rocket (which is analogous to the Hubble constant and is therefore known, at least in principle), but the mass of the Earth (or, in the cosmological case, the density of matter). It is therefore most useful to think about a critical density of matter, rather than a critical velocity. If the real density of matter exceeds the critical density, then the Universe will eventually recollapse: its gravitational energy is sufficient to slow down, stop, and then reverse the expansion. If the density is lower than this critical value, the Universe will carry on expanding forever. The critical density turns out to be extremely small. It also depends on H_0, but is on the order of one hydrogen atom per cubic metre. Most modern experimental physicists would consider material with such a low density to be a very good example of a vacuum!

And now, at last, we can introduce the quantity Ω: it is simply the ratio of the actual density of matter in the Universe to the critical value that marks the dividing line between eternal expansion and ultimate recollapse. $\Omega = 1$ marks that dividing line: $\Omega < 1$ means an ever-expanding Universe, and $\Omega > 1$ indicates one that recollapses in the future to a Big Crunch. Whatever the precise value of Ω, however, the effect of matter is always to slow down the expansion of the Universe, so that these models always predict a cosmic deceleration, but more of that shortly.

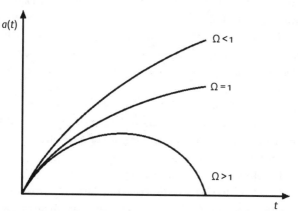

16. The Friedmann models. As well as having various options for curved space, the Friedmann models can also behave in different ways as they evolve with time. If Ω is greater than one then the expansion will eventually stop and the Universe will recollapse. If it is less than one the Universe will expand forever. Poised between these is the flat Universe with Ω finely tuned to be exactly unity.

But the long-term viability of the cosmological expansion is not the only issue whose resolution depends on Ω. These arguments based on simple ideas of energy resulting from Newtonian physics are not the whole story. In Einstein's general theory of relativity, the total energy-density of material determines the global curvature of space, as I described in Chapter 3. A space of negative global curvature results in models with Ω less than 1. A model with negative curvature is called an open universe model. A positively curved (closed) model pertains if Ω exceeds unity. In between, there is the classic British compromise universe, poised between eternal expansion and eventual recollapse, which has Ω exactly equal to unity. This model also has a flat geometry in which Euclid's theorems all apply. What a relief it would be if the Universe chose this simplest of all options!

The quantity Ω determines both the geometry of space on cosmological scales and the eventual fate of the Universe, but it is important to stress that the value of Ω is not at all *predicted* in the standard Big Bang model. It may seem to be a fairly useless kind of theory that is incapable of answering the basic questions that revolve around Ω, but in fact that is an unfair criticism. As I have explained, the Big Bang is a *model*, rather than a theory. As a model, it is self-consistent mathematically and compared to observations, but it is not complete. In this context this means that Ω is a 'free' parameter in much the same way as the Hubble constant H_o. To put it another way, the mathematical equations of the Big Bang theory describe the evolution of the Universe, but in order to calculate a specific example we need to supply a set of initial conditions to act as a starting point. Since the mathematics on which the model is based break down at the very beginning, we have no way of fixing the initial conditions theoretically. The Friedmann equation is well defined whatever the values of Ω and H_o, but our Universe happens to have been set up with one particular numerical combination of these quantities. All we can do, therefore, is to use observational data to make inferences about the cosmological parameters: they cannot, at least with the knowledge presently available and within the framework of the standard Big Bang, be deduced by reason alone. On the other hand, there is the opportunity to use present-day cosmological observations to learn about the very early Universe.

The search for two numbers

The importance of determining the cosmological parameters was recognized early on in the history of cosmology. Indeed, the distinguished astronomer Allan Sandage (formerly Hubble's research student) once wrote a paper entitled 'Cosmology: The Search for Two Numbers'. Two decades later, we still don't know the two numbers, and to understand why, we have to understand the different kinds of observation that can inform us about Ω, and what kind of results they have produced. There are many different types of observation, but they can be grouped into four main categories.

First, there are the classical cosmological tests. The idea with these tests is to use observations of very distant objects to measure the curvature of space, or the rate at which the expansion of the Universe is decelerating. The simplest of these tests involves the comparison of the ages of astronomical objects (particularly stars in globular cluster systems) with the age predicted by cosmological theory. I discussed this in Chapter 4 because if the expansion of the Universe were not decelerating, the predicted age depends much more sensitively on the Hubble constant than it does on Ω, and in any case the ages of old stars are not known with any great confidence, so this test is not a powerful diagnostic of Ω at the moment. Other classical tests involve using the properties of very distant sources to probe directly the rate of deceleration or the spatial geometry of the Universe. Some of these techniques were pioneered by Hubble and developed into an art form by Sandage. They fell into some disrepute in the 1960s and 1970s because it was then realized that not only was the Universe at large expanding, but objects within it were evolving rapidly. Since one needs to probe very large distances to measure the very slight geometrical effects of spatial curvature, one is inevitably looking at astronomical objects as they were when their light started out on its journey to us. This could be a very long time ago indeed: more than 80 per cent of the age of the Universe is commonplace in cosmological observations. There is no guarantee that the brightness or size of the distant objects being used had the same properties as nearby ones because of the possibility that these properties change with time. Indeed the classical cosmological tests are now largely used to study the evolution of properties, rather than to test fundamental aspects of cosmology. There is, however, one important and recent exception. The use of supernovae explosions as standard light sources has yielded spectacular results that seem to suggest the Universe is not decelerating at all. I'll talk more about these at the end of the chapter.

Next are arguments based on the theory of nucleosynthesis. As I explained in Chapter 5, the agreement between observed elemental abundances and the predictions of nuclear fusion calculations in the early Universe is one of the major pillars of evidence supporting the Big Bang theory. But this agreement only holds if the density of matter is very low indeed: no more than a few per cent of the critical density required to make space flat. This has been known for many years, and at first sight it seems to provide a very simple answer to all the questions I have posed. However, there is an important piece of small print attached to this argument. The 'few per cent' limit only applies to matter which can participate in nuclear reactions. The Universe could be filled with a background of sterile particles that were unable to influence the synthesis of the light elements. The kind of matter that involves itself in things nuclear is called *baryonic* matter and is made up of two basic particles, protons and neutrons. Particle physicists have suggested that other types of particle than baryonic ones might have been produced in the seething cauldron of the early Universe. At least some of these particles might have survived until now, and may make up at least some of the dark matter. At least some of the constituents of the Universe may therefore comprise some form of exotic non-baryonic particle.

Ordinary matter, of which we are made, may be but a small contaminating stain on the vast bulk of cosmic material whose nature is yet to be determined. This adds another dimension to the Copernican Principle: not only are we no longer at the centre of the cosmos, we're not even made from the same stuff as most of the Universe.

The third category of evidence is based on astrophysical arguments. The difference between these arguments and the intrinsically cosmological measurements discussed above is that they look at individual objects rather than the properties of the space between them. In effect, one is trying to determine the density of the Universe by weighing its constituent elements one by one. For example, one can attempt to use the internal dynamics of galaxies to work out their masses by assuming that the rotation of a galactic disk is maintained by gravity in much the same way as the motion of the Earth around the Sun is governed by the Sun's gravity. It is possible to calculate the mass of the Sun from the velocity of the Earth in its orbit, and a similar calculation can be done for galaxies: the orbital speeds of the stars in galaxies is determined by the total mass of the galaxy pulling on them. The principle can also be extended to clusters of galaxies, and systems of even larger size than this. These investigations overwhelmingly point to the existence of much more matter in galaxies than one sees there in the form of stars like our sun. This is the famous *dark matter* we can't see but whose existence we infer from its gravitational effects.

Rich clusters of galaxies – systems more than a million light years across consisting of huge agglomerations of galaxies – also contain more matter than is associated with the individual galaxies in them. The exact amount of matter is unclear, but there is very strong evidence that there is enough matter in the rich cluster systems to suggest that Omega is certainly as big as 0.1, and possibly even larger than 0.3. Tentative evidence from the dynamics of even larger structures – superclusters of clusters that are tens of millions of light years in size – suggests there

17. The Coma cluster. This is an example of a rich cluster of galaxies. Aside from the odd star (such as the one to the right of the frame), the objects in this picture are all galaxies contained within a giant cluster. Such enormous clusters are fairly rare but contain phenomenal amounts of mass, up to 100,000,000,000,000 times the mass of the Sun.

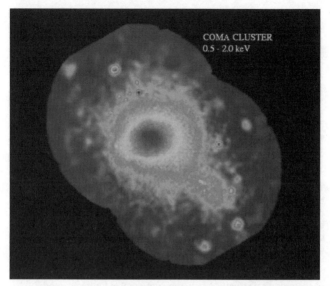

COMA CLUSTER
0.5 - 2.0 keV

18. Coma in X-rays. As well as the many hundreds of galaxies seen in the previous picture, clusters such as Coma also contain very hot gas that can be seen in the X-radiation it emits. This picture was taken by the ROSAT satellite.

may be even more dark matter lurking in the space between clusters. These dynamical arguments have also more recently been tested and confirmed against independent observations of the gravitational lensing produced by clusters, and by measurements of the properties of the very hot X-ray-emitting gas that pervades them. Intriguingly, the fraction of baryonic matter in clusters compared to their total mass seems much larger than the global value allowed by nucleosynthesis if there is a critical density of matter overall. This so-called *baryon catastrophe* means that either the overall density of matter is much lower than the critical value or some unknown process may have concentrated baryonic matter in clusters.

Finally, we have clues based on attempts to understand the origin

83

19. Gravitational lensing. Rich clusters can be weighed by observing the distortion of light from background galaxies as it passes through the cluster. In this beautiful example of the cluster Abell 2218 light from background sources is focused into a complicated pattern of arcs as the cluster acts as a giant lens. These features reveal the amount of mass contained within the cluster.

of cosmological structure: how the considerable lumpiness and irregularity of the Universe can have developed within a Universe that is required to be largely smooth by the Cosmological Principle. The idea behind how this is thought to happen in the Big Bang models is discussed in more detail in the next chapter. The basic principles are, I believe, relatively well understood. The details, however, turn out to be incredibly complicated and prone to all kinds of uncertainty and bias. Models can and have been constructed which seem to fit all the available data with Ω very close to unity. Others can do the same, with Ω much less than this. This may sound a bit depressing, but this kind of study probably ultimately holds the key to a successful determination of Ω. If more detailed measurements of the features in the microwave background can be made, then the properties of these features will tell us immediately what the density of matter must be. And, as a bonus, it will also determine the Hubble constant, bypassing all the tedious business of the cosmological distance ladder. We can only hope that the satellites planned to do this, MAP (NASA) and the Planck Surveyor (ESA), will fly successfully in the next few years. Recent balloon experiments have shown that this appears to be feasible, but I'll leave this to Chapter 7 to discuss further.

We can summarize the status of the evidence by suggesting that the vast majority of cosmologists probably accept that the value of Ω cannot be smaller than 0.2. Even this minimal value requires that most of the matter in the Universe is dark. It also means that at least some must not be in the form of protons and neutrons (baryons), which is where most of the mass resides in material with which we are familiar in everyday experience. In other words there must be non-baryonic dark matter. Many cosmologists favour a value of Ω around 0.3, which seems to be consistent with most of the observational evidence. Some have claimed that the evidence supports a value of the density close to the critical value, so that Ω can be very close to unity. This is partly because of the accumulating astronomical evidence for dark matter, but also because of the theoretical realization that non-baryonic matter might be produced at very high energies in the Big Bang.

The cosmic tightrope

The considerable controversy surrounding Ω is only partly caused by disagreements resulting from the difficulty of assessing the reliability and accuracy of (sometimes conflicting) observational evidence. The most vocal arguments in favour of a high value for Ω (i.e. close to unity) are based on theoretical, rather than observational, arguments. One might be inclined to dismiss such arguments as mere prejudice, but they have their roots in a deep mystery inherent to the standard Big Bang theory and which cosmologists take very seriously indeed.

To understand the nature of this mystery, imagine you are standing outside a sealed room. The contents of the room are hidden from you, except for a small window covered by a little door. You are told that you can open the door at any time you wish, but only once, and only briefly. You are told that the room is bare, except for a tightrope suspended in the middle about two metres in the air, and a man who, at some indeterminate time in the past began to walk the tightrope. You know also that if the man falls, he will stay on the floor until you open the door. If he doesn't fall, he will continue walking the tightrope until you look in.

What do you expect to see when you open the door? Whether you expect the man to be on the rope or on the ground depends on information you don't have. If he is a circus artist, he might well be able to walk to and fro along the rope for hours on end without falling. If, on the other hand, he were not a specialist in this area (like most of us), his stay on the rope would be relatively brief. One thing, however, is obvious. If the man falls, it will take him a very short time to fall from the rope to the floor. You would be very surprised, therefore, if your peep through the window happened to catch the man in transit from rope to ground. It is reasonable, on the grounds of what we know about this situation, to expect the man to be either on the rope or on the ground when we look, but if you see him in mid-tumble you would conclude that something fishy is going on.

This may not seem to have much to do with Ω, but the analogy becomes apparent with the realization that Ω does not have a constant value as time goes by. In the standard Friedmann models, Ω evolves and does so in a very peculiar way. At times arbitrarily close to the Big Bang, these models are all described by a value of Ω arbitrarily close to unity. To put this another way, look at Figure 16. Regardless of the behaviour at late times, all three curves shown get closer and closer together near the beginning and, in particular, they approach the 'flat Universe' line. As time passes, models with Ω just a little bit greater than unity in the early stages develop larger and larger values of Ω, with values far greater than unity when recollapse begins. Universes that start out with values of Ω just less than unity eventually expand much faster than the flat model, and then have values of Ω very close to zero. In the latter case, which is probably more relevant given the many indications that Ω is less than unity, the transition from Ω near unity, to Ω near zero is very rapid.

Now we can see the problem. If Ω is, say, 0.3, then in the very early stages of cosmic history it was very close to unity, but less than this value by a tiny amount. In fact, it really is a tiny amount indeed. At the Planck time, for example (i.e. 10^{-43} seconds after the Big Bang), Ω had to differ from unity only in the 60th decimal place. As time went by, Ω hovered close to the critical density state, only beginning to diverge rapidly in the recent past. In the very near future it will be extremely close to zero. But now, it is as if we caught the tightrope walker right in the middle of his fall. This seems very surprising, to put it mildly.

This paradox has become known as the Cosmological Flatness Problem, and it arises from the incompleteness of the standard Big Bang theory. That it is such a big problem convinced many scientists that it needed a big solution. The only way that seemed likely to resolve the conundrum was that our Universe really had to be a professional circus artist, to stretch the metaphor to breaking point. Obviously, Ω is not close to zero, as we have strong evidence of a lower limit to its value around 20 per cent. This rules out the man-on-the-ground alternative. The argument then goes that Ω must be equal to unity very closely, and that something must have happened in primordial times to single out this value very accurately.

Inflation and flatness

The happening that did this is claimed to be cosmological inflation, a speculation, originally made by Alan Guth in 1981, about the very early stages of the Big Bang model. Inflation involves a curious change in the properties of matter at very high energies known as a phase transition.

We have already come across an example of a phase transition. One occurs in the standard model about one-millionth of a second of the

Big Bang, and it involves the interactions between quarks. At low temperatures, quarks are confined in hadrons, whereas at higher temperatures they form a quark-gluon plasma. In between, there is a phase transition. In many unified theories, there can be many different phase transitions at even higher temperatures all marking changes in the form and properties of matter and energy in the Universe. Under certain circumstances a phase transition can be accompanied by the appearance of energy in empty space; this is called the vacuum energy. If this happens, the Universe begins to expand much more rapidly than it does in the standard Friedmann models. This is cosmic inflation.

Inflation has had a great impact on cosmological theory over the last twenty years. In this context, the most important thing about it is that the phase of extravagant expansion – which is very short-lived – actually reverses the way Ω would otherwise change with time. Ω is driven hard towards unity when inflation starts, rather than drifting away from it as it does in the cases described above. Inflation acts like a safety harness, pushing our tightrope walker back onto the wire whenever he seems like falling. An easy way of understanding how this happens is to exploit the connection I have established already between the value of Ω and the curvature of space. Remember that a flat space corresponds to a critical density, and therefore to a value of Ω equal to unity. If Ω differs from this magic value then space may be curved. If one takes a highly curved balloon and blows it up to an enormous size, say the size of the Earth, then its surface will appear flat. In inflationary cosmology, the balloon starts off a tiny fraction of a centimetre across and ends up larger than the entire observable Universe. If the theory of inflation is correct, then we should expect to be living in a Universe which is very flat indeed. On the other hand, even if Ω were to turn out to be very close to unity, that wouldn't necessarily prove that inflation happened. Some other mechanism, perhaps associated with quantum gravitational phenomena, might have trained our Universe to walk the tightrope.

These theoretical ideas are extremely important, but they cannot themselves decide the issue. Ultimately, whether theorists like it or not, we have to accept that cosmology has become an empirical science. We may have theoretical grounds for suspecting that Ω should be very close to unity, but observations must prevail in the end.

The sting

The question that emerges from all this is that if, as seems tentatively to be the case, Ω is significantly smaller than unity, do we have to abandon inflation? The answer is 'not necessarily'. For one thing, some models of inflation have been constructed that can produce an open, negatively curved Universe. Many cosmologists don't like these models, which do appear rather contrived. More importantly, there are now indications that the connection between Ω and the geometry of space may be less straightforward than has previously been thought. After many years in the wilderness, the classical cosmological tests I mentioned earlier have now staged a dramatic comeback. Two international teams of astronomers have been studying the properties of a particular type of exploding star, a Type Ia Supernova.

A supernova explosion marks the dramatic endpoint of the life of a massive star. Supernovae are among the most spectacular phenomena known to astronomy. They are more than a billion times brighter than the Sun and can outshine an entire galaxy for several weeks. Supernovae have been observed throughout recorded history. A supernova observed and recorded in 1054 gave rise to the Crab Nebula, a cloud of dust and debris inside which lies a rapidly rotating star called a pulsar. The great Danish astronomer Tycho Brahe observed a supernova in 1572. The last such event to be seen in our galaxy was recorded in 1604 and was known as Kepler's star. Although the average rate of these explosions in the Milky Way appears to be one or two every century or so, based on ancient records, none has been observed for nearly 400 years. In 1987, however, a supernova did explode in the Large Magellanic Cloud and was visible to the naked eye.

There are two distinct kinds of supernova, labelled Type I and Type II. Spectroscopic measurements reveal the presence of hydrogen in the Type II supernovae, but this is absent in the Type I versions. Type II supernovae are thought to originate directly from the explosions of massive stars in which the core of the star collapses to a kind of dead relic, while the outer shell is ejected into space. The final state of this explosion would be a neutron star or black hole. Type II supernova may result from the collapse of stars of different mass, so there is considerable variation in their properties from one to another. The Type I supernovae are further subdivided into Type Ia, Ib and Ic, depending on details of the shape of their spectra. The Type Ia supernovae are of particular interest. These have very uniform peak luminosities, for the reason that they are thought to be the result of the same kind of explosion. The usual model for these events is that a white dwarf star is gaining mass by accretion from a companion. When the mass of the white dwarf exceeds a critical mass called the *Chandrasekhar mass* (about 1.4 times the mass of the Sun), its outer parts explode while its central parts collapse. Since the mass involved in the explosion is always very close to this critical value, these objects are expected always to result in the liberation of the same amount of energy. The regularity of their properties means that Type Ia supernovae are very promising objects with which to perform tests of the curvature of space-time and the deceleration rate of the Universe.

New technology has enabled astronomers to search (and find) Type Ia in galaxies with redshifts around one. (Remember that this means that the Universe has expanded by a factor of two while light has been travelling from the supernova to us.) Comparing the observed brightness of the distant supernova with nearby ones can give an estimate of how much further away they are. This, in turn, tells us how much the Universe has been slowing down in the time taken for the light to reach us. The trouble is that these supernovae are fainter than they should be if the Universe is slowing down. The Universe is not decelerating at all, but speeding up.

This observation strikes at the heart of the standard description of cosmology embodied in the Friedmann equations. All these models should be decelerating. Even the members of the Friedmann family with low Ω, whose deceleration is only slight owing to their low density, should not be speeding up. And the models with critical density apparently favoured by inflation should undergo heavy deceleration. What has gone wrong?

Einstein's biggest wonder?

The supernova observations I've been talking about are still controversial, but they certainly seem to indicate that a dramatic change in cosmological theory is needed. On the other hand, there is an off-the-shelf remedy for this bug that dates from Einstein himself. In Chapter 3 I mentioned how Einstein altered his original theory of gravitation by introducing a cosmological constant. His reason for doing this, an act he later regretted, was that he wanted to make a theory that could describe a static (i.e. non-expanding) universe. His cosmological constant altered the law of gravity to prevent space from either expanding or contracting. Applied in the modern context, the cosmological constant can be introduced to make the law of gravity repulsive on large scales. If this is done, the tendency of the gravitational attraction of matter to slow down the Universe is overwhelmed by a cosmic repulsion that causes it to speed up.

This cure of course requires one to accept that the cosmological constant wasn't a bad idea in the first place. But modern theory also gives us a new understanding of how this can happen. In Einstein's original theory, the cosmological constant appeared in the mathematical equations describing gravity and space-time curvature.

It was indeed a modification of the law of gravity. But he could just as easily have written this term on the other side of his equations, in the part of the theory that describes matter. Over on the other side of Einstein's equations, his infamous cosmological constant appears as a term describing the energy density of the vacuum. A vacuum that has energy may sound strange, but we have come across it before in this chapter. It is exactly what is needed to cause inflation.

In the early versions of cosmic inflation theory, the vacuum energy liberated by a primordial phase transition disappears after the transient period of hyper-expansion is over. But maybe a small amount of this energy survived until now and it is this energy that has made gravity push instead of pull. The idea that this vacuum energy may be causing the acceleration also allows us to reconcile the theory of inflation with the evidence that Ω may be significantly less than the unit value it would have to have if space were flat. While the vacuum energy is perverse in that it makes gravity push rather than pull, it does at least curve space in the same way that ordinary matter does. If we have a Universe with both matter and vacuum energy then we can have flat space without the deceleration that is required in an ordinary Friedmann model.

We still don't really know for sure whether the Universe is accelerating, whether there is a vacuum energy, or what precisely is the value of Ω. But these ideas have provoked intense activity over the last few years in both theory and experiment. And there is a new generation of measurements coming along that could, if they work, answer all these questions. I'll discuss these in the next chapter.

Chapter 7
Cosmic structures

Galaxies are the basic building blocks of the Universe. They are not, however, the largest structures one can see. They tend not to be isolated, but like to band together rather like people. The term used to describe the way galaxies are distributed over cosmological distances is *large-scale structure*. The origin of this structure is one of the hot topics of modern cosmology but, before explaining why this is so, it is first necessary to describe what the structure actually is.

Patterns in space

The distribution of matter on large scales is usually determined by means of spectroscopic surveys that use Hubble's Law to estimate the distances to galaxies from their redshifts. The existence of structure was known for many years before redshift surveys became practicable. The distribution of galaxies on the sky is highly non-uniform, as can be seen in the first large systematic survey of galaxy positions which resulted in the *Lick Map*. But impressive though this map undoubtedly is, one cannot be sure if the structures seen in it are real, physical structures or just chance projection effects. After all, we all recognize the constellations, but these are not physical associations. The stars in them lie at very different distances from the Sun. For this reason, the principal tool of cosmography has become the redshift survey.

20. The Andromeda Nebula. The nearest large spiral galaxy to the Milky Way, Andromeda is a good example of its type. Not all galaxies are spiral; rich clusters like Coma contain mainly elliptical galaxies with no spiral arms.

A famous example of this approach is the Harvard-Smithsonian Center for Astrophysics (CfA) survey, which published its first results in 1986. This was a survey of the redshifts of 1,061 galaxies found in a narrow strip on the sky in the original Palomar Sky Survey published in 1961. This survey has subsequently been extended to several more strips by the same team. Until the 1990s redshift surveys were slow and laborious because it was necessary to point a telescope at each galaxy in turn, take a spectrum, calculate the redshift and then move to the next galaxy. To acquire several thousands of redshifts took months of telescope time, which, because of the competition for resources, would usually be spread over several years. More recently the invention of multi-fibre devices on wide-field telescopes has allowed astronomers to capture as many as 400 spectra in one pointing of the telescope. Among the latest generation of redshift surveys is one called the Two-Degree Field (2dF) survey, run by the United Kingdom and Australia using the Anglo-Australian

Telescope. This will eventually map the positions of around 250,000 galaxies.

The general term used to describe a physical aggregation of many galaxies is a *cluster of galaxies*, or *galaxy cluster*. Clusters can be systems of greatly varying size and richness. For example, our galaxy, the Milky Way, is a member of the so-called *Local Group* of galaxies, which is a rather small cluster of galaxies of which the only other large member is the Andromeda galaxy (M31). At the other extreme, there are the so-called *rich clusters of galaxies*, also known as *Abell clusters*, which

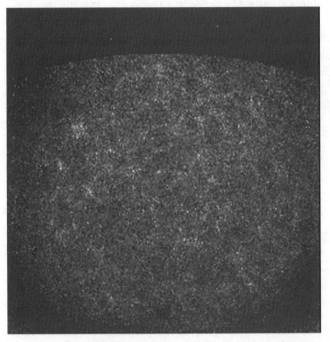

Cosmic structures

21. The *Lick Map*. Produced by meticulous eyeball counting of galaxies on survey plates, the *Lick Map* displays the distribution of about a million galaxies over the sky. The pattern of filaments and clusters is impressive; the dense round lump near the centre is the Coma cluster.

contain many hundreds or even thousands of galaxies in a region just a few million light years across: prominent nearby examples of such entities are the Virgo and Coma clusters. In between these two extremes, galaxies appear to be distributed in systems of varying density in a roughly fractal (or hierarchical) manner. The densest Abell clusters are clearly collapsed objects held together in equilibrium by their own self-gravity. The less rich and more spatially extended systems may not be bound in this way, but may simply reflect a general statistical tendency of galaxies to clump together.

Individual galaxy clusters are still not the largest structures to be seen. The distribution of galaxies on scales larger than around 30 million light years also reveals a wealth of complexity. Recent observational surveys have shown that galaxies are not simply distributed in quasi-spherical 'blobs', like the Abell clusters, but also sometimes lie in extended quasi-linear structures called *filaments*, or flattened sheet-like structures such as the *Great Wall*. This is a roughly two-dimensional concentration of galaxies, discovered in 1988 by astronomers from the Harvard-Smithsonian Center for Astrophysics. The Great Wall is at least 200 million light years by 600 million light years in size, but is less than 20 million light years thick. It contains many thousands of galaxies and has a mass of at least 10^{16} solar masses. The rich clusters themselves are clustered into enormous loosely bound agglomerations called *superclusters*. Many are known, containing anything from around ten rich clusters to more than fifty. The most prominent known supercluster is called the *Shapley concentration*, while the nearest is the Local Supercluster, centred on the Virgo cluster mentioned above, a flattened structure in the plane of which the Local Group is moving. Superclusters are known with sizes as large as 300 million light years, containing as much as 10^{17} solar masses of material.

These structures are complemented by vast nearly empty regions, many of which appear to be roughly spherical. These 'voids' contain very many fewer galaxies than average, or even no galaxies at all. Voids with

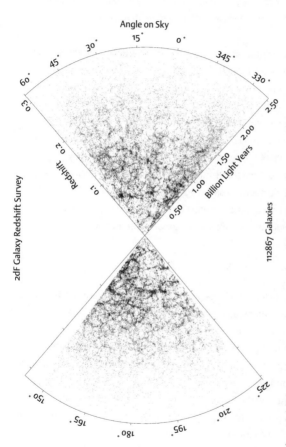

22. The 2dF galaxy redshift survey. This survey, which is still in progress, is planned to measure the redshifts of around 250,000 galaxies. Although parts of the survey are not finished, resulting in missing pieces of the map, one can see the emergence of a complex network of structures extending out to billions of light years from us.

density less than 10 per cent of the average density on scales of up to 200 million light years have been detected in large-scale redshift surveys. The existence of large voids is not surprising, given the existence of clusters of galaxies and superclusters on very large scales, because it is necessary to create regions of less than average density for there to be regions of greater than average density.

The impression one has when looking at maps of large-scale structure is that of a vast cosmic 'web', a complex network of intersecting chains and sheets. But how did this complexity arise? The Big Bang model is predicated on the assumption that the Universe is smooth and featureless, i.e. that it conforms to the Cosmological Principle. Fortunately the structure does indeed seem to peter out on scales larger than the scale of the cosmic mesh. This is also confirmed by observations of the cosmic microwave background, which comes to us after travelling about 15 billion light years from the early Universe. The microwave background is almost uniform on the sky, consistent with the Cosmological Principle. Almost, that is, but not quite.

Structure formation

In 1992, the COBE satellite deployed its sensitive detectors to the task of detecting and mapping any variation in the temperature of the microwave background on the sky. At its discovery in 1965 the microwave background seemed to be isotropic on the sky. Later it was found that it had a large-scale variation across the sky of about one part in a thousand of the temperature. This is now known to be a Doppler effect, caused by the Earth's motion through the radiation field left over from the Big Bang. The sky looks slightly warmer in the direction we are heading, and slightly cooler in the direction we are coming from. But aside from this 'dipole' variation (as it is called), the radiation seemed to be coming equally from all directions. But theorists had suspected for a long time that there should be structure in the microwave background, in the form of a ripply pattern of hot and cold splotches. It was these that COBE found, and in so doing, caused newspaper headlines around the world.

So why is the microwave background not smooth after all? The answer is intimately connected to the origin of large-scale structure and, as ever in cosmology, gravity provides the connection.

Friedmann's models provide important insights into how the bulk properties of the Universe change with time. But they are unrealistic because they describe an idealized world that is perfectly smooth and blemish-free. A Universe that starts out like that will remain perfect forever. In a realistic situation, however, there are always imperfections. Some regions may be slightly denser than average, some more rarefied. How does a slightly lumpy Universe behave? The answer is dramatically different to the idealized case. A piece of the Universe that is denser than average exerts a stronger gravitational pull on its surroundings than average. It will therefore tend to suck material in, depleting its neighbourhood. In the process it gets even denser relative to the average, and pulls still harder. The effect is a runaway growth of lumpiness called the 'gravitational instability'. Eventually strongly bound lumps form and begin to collect into filaments and sheets resembling those seen in maps of cosmic structure. Only very slight fluctuations are needed to kick the process off, but gravity acts like a powerful amplifier transforming minute initial ripples into huge fluctuations in density. We can map the end product using galaxy surveys; we see the initial input in the COBE map. We even have a good theory of how the initial fluctuations imprinted; cosmic inflation produces quantum fluctuations.

The basic picture of how structure forms has been around for many years, but it is hard to turn this into detailed predictive calculations because of the complicated behaviour of gravity. I mentioned in Chapter 3 that even Newton's laws of motion are difficult to solve without simplifying symmetry. In the late stages of gravitational instability, there is no such simplification. Everything in the Universe pulls on

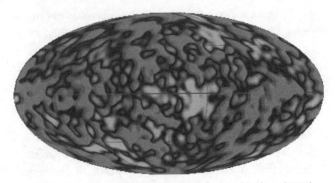

23. The COBE ripples. In 1992 the Cosmic Background Explorer (COBE) satellite measured slight fluctuations of about one part in 100,000 of the temperature of the cosmic microwave background on the sky. These 'ripples' are thought to be the seeds from which galaxies and large-scale structure grew.

everything else; it is necessary to keep track of all the forces acting everywhere and on everything. The sums involved are just too hard to be solved with pencil and paper.

During the 1980s, however, massive computers came on the scene, and progress in the field accelerated. It became obvious that gravity could form cosmic structure but in order for it to do the job effectively there would have to be quite a lot of mass in the Universe. Because only a relatively small amount of 'normal' matter is allowed by primordial nucleosynthesis arguments, theorists assumed the Universe to be dominated by some form of exotic dark matter that does not involve itself in nuclear reactions. Simulations showed that the best form of matter for this was 'cold' dark matter. If the dark matter were 'hot' then it would be moving too quickly to form clumps of the right size.

Eventually, after many years of computer time, a picture emerged in which cosmic structure arises in a bottom-up fashion. First, small clumps of dark matter form. These building blocks then coalesce into larger units, which then themselves coalesce, and so on. Eventually objects the size of galaxies form. Gas (which is made of baryonic material) falls in, stars form, and we have galaxies. The galaxies continue the hierarchical growth of structure by clustering in chains and sheets. In this picture, structure evolves rapidly with time (or, equivalently, with redshift).

The idea of cold dark matter has been very successful, but this programme is far from complete. It is still not known how much dark matter there is, nor what form it takes. The detailed problem of how

24. The Hubble deep field. Made by pointing the Hubble Space Telescope at a blank piece of sky, this image shows a wonderful array of distant faint galaxies. Some of these objects are at such enormous distances that light has taken more than 90 per cent of the age of the Universe to reach us. We can therefore see galaxy evolution happening.

galaxies form is also unsolved because of the complex hydrodynamical and radiative processes involved with the motion of gas and the formation of stars. But now this field is not just about theory and simulations. Breakthroughs in observational technology, such as the Hubble Space Telescope, now allow us to see galaxies at high redshift and thus study precisely how their properties and distribution in space has changed with time. With the next generation of huge redshift surveys we will have enormously detailed information about the pattern that galaxies trace out in space. This too holds clues as to how much dark matter there is, and precisely how galaxies formed. But the final resolution of this problem is likely to come not from observations of the end product of the gravitational instability process but at its beginning.

The sound of creation

The COBE satellite represented an enormous advance in the study of structure formation, but in many ways this experiment was very limited. The most important shortcoming of COBE was that it lacked the ability to resolve the detailed structure of the ripples in the microwave background. In fact COBE's angular resolution was only about ten degrees, which is very crude by astronomical standards. The full Moon, for comparison, is about half a degree across. It is in the fine structure of the microwave sky that cosmologists hope to find the answers to many outstanding questions.

The ripples in the early Universe were produced by a kind of sound wave. When the Universe was very hot, with a temperature of several thousand degrees, it was ringing with sound waves travelling backwards and forwards. The surface of the Sun is at a similar temperature and is vibrating in a similar way. Because of its poor resolution COBE was able to detect only those ripples that have a very long wavelength. These represent sound waves of very low pitch, the bass notes of creation. The information contained in these waves is important but not very detailed; their sound is rather dull.

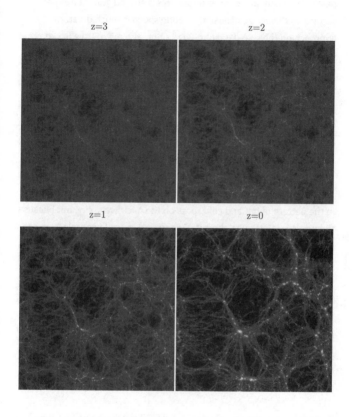

z=3　　　z=2

z=1　　　z=0

25. Simulation of structure formation. Starting from almost smooth initial conditions, modern supercomputers can be used to evolve a simulated chunk of the Universe forward in time. In this example, performed by the Virgo Consortium, we can see hierarchical clustering develop as the Universe expands by a factor of 4. The dense knots seen in the last frame form galaxies and galaxy clusters, while the filamentary structure is strongly reminiscent of that seen in the galaxy surveys.

On the other hand, the Universe should also produce sound of higher pitch and this is much more interesting. Sound waves travel with a particular speed. In air, for example, this is around 300 metres per second. In the early Universe the sound speed is much greater, approaching the speed of light. By the time the microwave background is produced the Universe is about 300,000 years old. In the time up to then since the Big Bang, which is presumably when the sound waves were excited in the first place, they can have travelled only about 300,000 light years. Oscillations with this wavelength produce a characteristic 'note', like the fundamental tone of a musical instrument. It is no coincidence that superclusters of galaxies are roughly of this size; they result from this resounding cosmic fanfare.

The characteristic wavelength of the early Universe should reveal itself in the pattern of hot and cold spots on the microwave sky, but because the wavelength is quite short it appears as a much finer scale than can be resolved by COBE. In fact, the angular size of the spots it produces is around one degree. Since COBE, therefore, there has been a race to develop instruments capable of detecting not just the fundamental tone of the Universe but also its higher harmonics. By a detailed analysis of the sound of creation, it is hoped to answer many of the major questions facing modern cosmology. The spectrum of sound contains information about how much mass there is, whether there is a cosmological constant, what the Hubble constant is, whether space is curved, and perhaps even whether inflation happened or not.

Two major experiments, the NASA-led MAP mission to be launched in 2001 and the European Space Agency's Planck Surveyor to be launched a few years later, will make detailed maps of the pattern of ripples on the sky with very high resolution. If the interpretation of these structures is correct, we should have definite answers very soon. The sense of anticipation is palpable.

In the mean time, there are very strong hints as to how things will pan out. Two important balloon-borne experiments, BOOMERANG and MAXIMA, have mapped little bits of the sky with only slightly poorer resolution than MAP and Planck will. These experiments have not yielded definitive answers, but they do indicate that the geometry of the Universe is flat. The argument is simple. We know the characteristic wavelength of the sounds producing the measured features. We know how far away these waves are observed (about 15 billion light years). We can therefore work out what angle they should occupy on the sky if the Universe is flat. If the Universe is open the angle will be smaller than it would be in a flat Universe; if it is closed the angle will be larger. The results imply flatness. Together with the acceleration I talked about in the previous chapter, this measurement also provides strong evidence

26. BOOMERANG. The picture shows this experiment about to be launched on a balloon in Antarctica. The experimental payload is perched on the vehicle to the right. The flight path of this balloon took it around the South Pole, making use of circulating winds to return it near the point of launch. Antarctica is very dry, making it the best place on Earth for microwave background experiments, but it is still better to get out into space if possible.

for a cosmological constant. The only way we know of having a flat yet accelerating cosmos is if there is vacuum energy.

The picture emerging from structure studies seems to be in line with the other strands I have discussed, but we still don't know how the Universe contrived to be the way it is. The answer to this deeper puzzle will rely on deeper understandings of the nature of matter, space, and time. I'll discuss these in the next chapter.

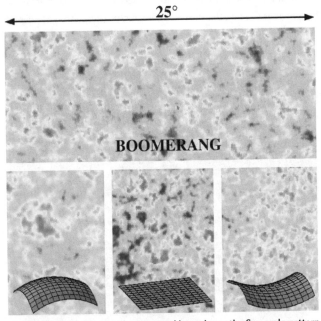

27. The flatness of space. The top panel here shows the fine-scale pattern of temperature fluctuations measured by BOOMERANG. Below are simulated patterns that take into account the expected angular size of these fluctuations in closed, flat, and open cosmologies. The best match is with a flat Universe (centre). This strong indication has added impetus to the future experiments MAP and the Planck Surveyor that will map the whole sky with this resolution.

Chapter 8
A theory of everything?

The modern era of physics began with two revolutions that took place
in the early years of the twentieth century. One of them involved the
introduction of relativity, and it played a major role in the development
of cosmology throughout this century. The other major upheaval was
the birth of quantum mechanics. By contrast, the implications of
quantum physics for cosmology are still far from understood.

The world of the quantum

In the world according to quantum theory, every entity has a dual
nature. In classical physics two distinct concepts were used to describe
distinct natural phenomena: waves and particles. Quantum physics tells
us that these concepts do not apply separately to the microscopic
world. Things that we previously imagined to be particles can
sometimes behave like waves. Phenomena that we previously thought
of as waves can sometimes behave like particles. Light behaves like a
wave. One can produce interference and diffraction effects using prisms
and lenses. Moreover, Maxwell had shown that light was actually
described mathematically by an equation called the wave equation: the
wave nature of light is therefore predicted by this theory. On the other
hand, Max Planck's work on the radiation emitted by hot bodies had
also shown that light could also behave as if it came in discrete packets,
which he called *quanta*. He hesitated to claim that these quanta could

be identified with particles. It was in fact Albert Einstein, in the work on the photoelectric effect for which he won the Nobel Prize, who made the step of saying that light was actually made of particles. These particles later became known as *photons*. But how can something be both a wave and a particle? One has to say that reality cannot be exactly described by either concept, but that it behaves sometimes as if it were a wave and sometimes as if it were a particle.

Imagine a medieval monk returning to his monastery after his first trip to Africa. During his travels he chanced upon a rhinoceros, and is faced with the task of describing it to his incredulous brothers. Since none of them has ever seen anything as strange as a rhino in the flesh, he has to proceed by analogy. The rhinoceros, he says, is in some respects like a dragon and in others like a unicorn. The brothers then have a reasonable picture of what the beast looks like. But neither dragons nor unicorns exist in nature, while the rhinoceros does. It is the same with our quantum world: reality is described neither by idealized waves nor by idealized particles, but these concepts can give some impression of certain aspects of the way things really are.

The idea that energy came in discrete packets (or quanta) was also successfully applied to the simplest of all atoms, the hydrogen atom, by Niels Bohr in 1913 and to other aspects of atomic and nuclear physics. The existence of discrete energy levels in atoms and molecules is fundamental to the field of spectroscopy, which plays a role in fields as diverse as astrophysics and forensic science and was crucial to Hubble's discovery of the recession of the galaxies.

The uncertain Universe

The acceptance of the quantized nature of energy (and light) was only the start of the revolution that founded modern quantum mechanics. It was not until the 1920s and the work of Schrödinger and Heisenberg that the dual nature of light as both particle and wave was finally elucidated. For while the existence of photons had become accepted in the previous years, there had been no way to reconcile this with the well-known wave behaviour of light. What emerged in the 1920s was a theory of quantum physics built upon wave mechanics. In Schrödinger's version of quantum theory, the behaviour of all systems is described in terms of a wavefunction (usually called ψ), which evolves according to an equation called the Schrödinger equation. The wavefunction ψ depends on both space and time. Schrödinger's equation describes waves that fluctuate in both space and time.

So how does the particle behaviour come in? The answer is that the quantum wavefunction does not describe something like an electromagnetic wave, which one thinks of as a physical thing existing at a point in space and fluctuating in time. The quantum wavefunction describes a 'probability wave'. Quantum theory asserts that the wavefunction is all one can know about the system: one cannot predict with certainty exactly where the particle will be at a given time, just the probability.

An important aspect of this wave-particle duality is the Uncertainty Principle. This has many repercussions for physics, but the simplest one involves the position of a particle and its speed. Heisenberg's uncertainty principle states that one cannot know the position and speed of a particle independently of one another. The better you know the position, the worse you know the speed, and vice versa. If you can pinpoint the particle exactly, then its speed is completely unknown. If you know its speed precisely, then the particle could be located anywhere. This principle is quantitative, does not apply only to position and momentum, but also to energy and time and other pairs of quantities that are known as conjugate variables.

It is a particularly important consequence of the energy-time Uncertainty Principle that empty space can give birth to short-lived particles that spring in and out of existence on a timescale controlled by the Uncertainty Principle. This is the reason why particle physicists expect the vacuum to possess energy. In other words, there should be a cosmological constant. The only problem is that they don't know how to calculate it. The best guesses available are too large by more than 100 orders of magnitude. But the idea of cosmic uncertainty has scored one notable success: it is thought to be the reason for the existence of small primordial density fluctuations that started off the growth of cosmic structure.

A Universe running according to Newtonian physics is *deterministic*, in the sense that if one knew the positions and velocities of all the particles in a system at a given time then one could predict their behaviour at all subsequent times. Quantum mechanics changed all that, since one of the essential components of this theory is the principle that at a fundamental level, the behaviour of particles is inherently unpredictable, hence the need to resort to calculations of probability.

The interpretation to be put on this probabilistic approach is open to considerable debate. For example, consider a system in which particles travel in a beam towards two closely separated slits. The wavefunction ψ corresponding to this situation displays an interference pattern because the 'probability wave' passes through both slits. If the beam is powerful, it will consist of huge numbers of photons. Statistically the photons should land on a screen behind the slits according to the probability dictated by the wavefunction. Since the slits set up an interference pattern, the screen will show a complicated series of bright and faint bands where the waves sometimes add up in phase and sometimes cancel each other. This seems reasonable, but suppose we turn the beam down in power. This can be done in such a way that there is only one photon at any time travelling through the slits. The arrival of each photon can be detected on the screen. By running the experiment for a reasonably long time one can build up a pattern on the screen. Despite the fact that only one photon at a time is travelling through the apparatus, the screen still shows the pattern of fringes. In some sense each photon must turn into a wave when it leaves the source, travel through both slits, interfering with itself on the way, and then turn back into a photon in order to land in a definite position on the screen.

So what is going on? Clearly each photon lands in a particular place on the screen. At this point we know its position for sure. What does the wavefunction for this particle do at this point? According to one interpretation – the so-called Copenhagen interpretation – the wavefunction collapses so that it is concentrated at a single point. This happens whenever an experiment is performed and a definite result is obtained. But before the outcome is settled nature itself is indeterminate: the photon really doesn't go through either one of the slits: it is in a 'mixed' state. The act of measurement changes the wavefunction and therefore changes reality. This has led many to speculate about the interaction between consciousness and quantum 'reality'. Is it consciousness that causes the wavefunction to collapse?

A famous illustration of this conundrum is provided by the paradox of *Schrödinger's Cat*. Imagine there is a cat inside a sealed room containing a vial of poison. The vial is attached to a device which will break it and poison the cat when a quantum event occurs, for example the emission of an alpha-particle by a lump of radioactive material. If the vial breaks, death is instantaneous. Most of us would accept that the cat is either alive or dead at a given time. But if one takes the Copenhagen interpretation seriously it is somehow both: the wavefunction for the cat comprises a superposition of the two possible states. Only when the room is opened and the state of the cat 'measured' does it 'become' either alive or dead.

An alternative to the Copenhagen interpretation is that nothing physically changes at all when a measurement is performed. What happens is that the observer's state of knowledge changes. If one asserts that the wavefunction ψ represents what is known by the observer rather than what is true in reality then there is no problem in having it change when a particle is known to be in a definite state. This view suggests an interpretation of quantum mechanics in which at some level things might be deterministic, but we simply do not know enough to predict.

Yet another view is the Many Worlds interpretation. In this, every time an experiment is performed (e.g. every time a photon passes through the slit device) the universe, as it were, splits into two: in one universe the photon goes through the left-hand slit and in the other it goes through the right-hand slit. If this happens for every photon one ends up with an enormous number of parallel universes. All possible outcomes of all possible experiments occur in this ensemble. But before I head off into a parallel universe, let me resume the thread of the story.

The missing link

I described the standard model of fundamental interactions in Chapter 5. The three forces it incorporates are all described by quantum theories. The fourth of the fundamental interactions is gravity. This has proved extremely resistant to efforts to make it fit into a unified scheme of things. The first step in doing so would involve incorporating quantum physics into the theory of gravity in order to produce a theory of quantum gravity. Despite strenuous efforts, this has not yet been achieved. If this is ever done, the next task will be to unify quantum gravity with a unified theory of the particle interactions.

It is ironic that it is general relativity, which really began the modern era of theoretical physics, that should provide the stumbling block to further progress towards a unified theory of all the forces of nature. In many ways, the force of gravity is extremely weak. Most material bodies are held together by electrical forces between atoms which are many orders of magnitude stronger than the gravitational forces between

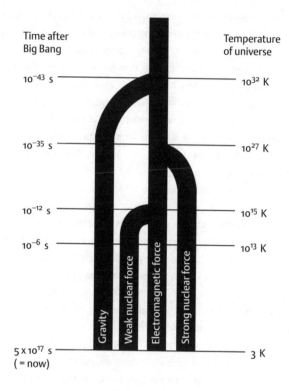

Time after
Big Bang

Temperature
of universe

10^{-43} s 10^{32} K

10^{-35} s 10^{27} K

10^{-12} s 10^{15} K

10^{-6} s 10^{13} K

Gravity

Weak nuclear force

Electromagnetic force

Strong nuclear force

5×10^{17} s   3 K
(= now)

28. A theory of everything. The four forces of Nature we know in our low-energy world are thought to become inextricably unified at higher energy. Turning the clock back on the Big Bang we first expect that electromagnetism and weak interactions merge into an electroweak force. At higher energies still, this electroweak force will unite with the strong nuclear force in a Grand Unified Theory (GUT). At higher energies still, gravity may join in to produce a theory of everything. It is this theory, if it exists, that will describe the Big Bang itself.

them. But despite its weakness gravity has a perplexing nature that seems to resist attempts to put it together with quantum theory.

Einstein's general theory of relativity is a classical theory, in the sense that Maxwell's equations of electromagnetism are also classical; they involve entities that are smooth rather than discrete and describe behaviour that is deterministic rather than probabilistic. On the other hand, quantum physics describes a fundamental lumpiness: everything consists of discrete packets or quanta. Likewise, the equations of general relativity allow one to calculate the exact state of the Universe at a given time in the future if sufficient information is given at some time in the past. They are therefore deterministic. The quantum world, on the other hand, is subject to the uncertainty embodied in Heisenberg's Uncertainty Principle.

Of course, classical electromagnetic theory is perfectly adequate for many purposes, but the theory does break down in certain situations, such as when radiation fields are very strong. For this reason physicists sought (and eventually found) the quantum theory of electromagnetism or quantum electrodynamics (QED). This theory was also made consistent with the special theory of relativity, but does not include general-relativistic effects.

While Einstein's equations also seem quite accurate for most purposes, it is similarly natural to attempt the construction of a quantum theory of gravity. Einstein himself always believed that his theory was incomplete in this sense, and would eventually need to be replaced by a more complete theory. By analogy with the breakdown of classical electromagnetism, one can argue that this should happen when gravitational fields are very strong, or on length scales that are extremely short. Attempts to build such a theory have so far been unsuccessful.

Although there is nothing resembling a complete picture of what a quantum theory of gravity might involve, there are some interesting speculative ideas. For example, since general relativity is essentially a theory of space-time, space and time themselves must become quantized in quantum gravity theories. This suggests that, although space and time appear continuous and smooth to us, on minuscule scales of the Planck length (around 10^{-33} cm), space is much more lumpy and complicated, perhaps consisting of a foam-like topology of bubbles connected by tunnels called wormholes that are continually forming and closing in the Planck time, which is 10^{-43} seconds. It also seems to make sense to imagine that quantized gravitational waves, or *gravitons*, might play the role of the gauge bosons in other fundamental interactions, such as the photons in the theory of quantum electrodynamics. As yet, there is no concrete evidence that these ideas are correct.

The tiny scales of length and time involved in quantum gravity demonstrate why this quantum gravity is a field for theorists rather than experimentalists. No device has yet been built capable of forcing particles into a region of order the Planck length or less. The enormous energies required to do this are needed to reveal the quantum nature of gravity. But that is precisely why so many theoreticians have turned away from particle experiments as such, and towards cosmology. The Big Bang must have involved phenomena on the Planck scale, so it may in principle be possible to learn about fundamental physics from cosmology.

The beginning of time

The presence of a singularity at the very beginning of the Universe is very bad news for the Big Bang model. Like the black hole singularity, it is a real singularity where the temperature and density become truly infinite. In this respect the Big Bang can be thought of as a kind of time-reverse of the gravitational collapse that forms a black hole. As was the case with the Schwarzschild solution, many physicists thought

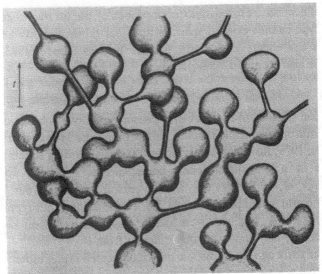

29. Space-time foam. One of the ideas associated with quantum gravity is that space-time itself may turn into a seething mass of bubbles and tubes all springing into existence and disappearing on a timescale comparable to the Planck time.

that the initial cosmological singularity could be a consequence of the special form of the solutions of Einstein's equations used to model the Big Bang, but this is now known not to be the case. Hawking and Penrose generalized Penrose's original black hole theorems to show that a singularity invariably exists in the past of an expanding Universe in which certain very general conditions apply. Physical theory completely fails us at the instant of the Big Bang where the nasty infinities appear.

So is it possible to avoid this singularity? And if so, how? It is most likely that the initial cosmological singularity might well just be a consequence of extrapolating deductions based on the classical theory of general relativity into a situation where this theory is no longer valid.

This is what Einstein says in the paragraph quoted in Chapter 3 during the discussion of black holes. What is needed is quantum gravity, but we don't have such a theory and, since we don't have it we don't know whether it would solve the riddle of the Universe's apparently pathological birth.

There are, however, ways of avoiding the initial singularity in classical general relativity without appealing to quantum effects. Firstly, one could try to avoid the singularity by proposing an equation of state for matter in the very early Universe that does not obey the conditions laid down by Hawking and Penrose. The most important of these conditions is a restriction on the behaviour of matter at high energies called the *strong energy condition*. There are various ways in which this condition might indeed be violated. In particular, it is violated during the accelerated expansion predicted in theories of cosmic inflation. Models in which this condition is violated right at the very beginning can have a 'bounce' rather than a singularity. Running the clock back, the Universe reaches a minimum size and then expands again.

Whether the singularity is avoidable or not remains an open question, and the issue of whether we can describe the very earliest phases of the Big Bang, before the Planck time, will remain open at least until a complete theory of quantum gravity is constructed.

Time's arrow

The existence of a singularity at the beginning of the Universe calls into question the very nature of space, and particularly of time, at the instant of creation. It would be nice to include at this point a clear definition of what time actually is. Everyone is familiar with what time does, and how events tend to be ordered in sequences. We are used to describing events that invariably follow other events in terms of a chain of cause and effect. But we can't get much further than these simple ideas. In the end, the best statement of what is time is that time is whatever it is that is measured by clocks.

Einstein's theories of relativity effectively destroyed the Newtonian concepts of absolute space and absolute time. Instead of having three spatial dimensions and one time dimension which are absolute and unchanging regardless of the motions of particles or experimenters, relativistic physics merges these together in a single four-dimensional entity called space-time. For many purposes, time and space can be treated as mathematically equivalent in these theories: different observers generally measure different time intervals between the same two events, but the four-dimensional space-time interval is always the same.

However, the successes of Einstein's theoretical breakthroughs tend to mask the fact that we all know from everyday experience that time and space are essentially different. We can travel north or south, east and west, but we can only go forwards in time to the future, not backwards in time to the past. And we are quite happy with the idea that both London and New York exist at a given time at different spatial locations. But nobody would say that the year 5001 exists in the same way that we think the present exists. We are also happy to say that what we do now causes things to happen in the future, but do not consider events at the same time in two locations as causing each other. Space and time really are quite different.

On a cosmological level, the Big Bang certainly appears to have a preferred direction. But the equations describing it are again time-symmetric. Our universe happens to be expanding rather than contracting, but it could have been collapsing and described by the same laws. Or could it be that the directionality of time that we observe is somehow singled out by the large-scale expansion of the Universe? It has been speculated, by Hawking and others, that if we lived in a closed Universe that eventually stopped expanding and began to contract, then time would effectively run backwards during the contraction phase. In fact, if this happened we would not be able to tell the difference between a contracting Universe with time running backwards and an expanding Universe with time running forwards. Hawking was convinced for a time that this had to be the case, but later changed his mind.

A more abstract problem stems from the fact that Einstein's theory is fully four-dimensional: the entire world-line of a particle, charting the whole history of its motions in space-time, can be calculated from the theory. A particle which exists at different times exists in the same way two particles might exist at the same time in different places. This is strongly at odds with our ideas of free will. Does our future exist already? Are things really predetermined in this way?

These questions are not restricted to relativity theory and cosmology. Many physical theories are symmetric between past and future in the same way as they are symmetric between different spatial locations. The question of how the perceived asymmetry of time can be reconciled with these theories is a deep philosophical puzzle. There are at least two other branches of physical theory in which raise the question of the *arrow of time*, as it is sometimes called.

One emerges directly from a seemingly omnipotent physical principle, called the Second Law of Thermodynamics. This states that the *entropy* of a closed system never decreases. The entropy is a measure of the disorder of a system, so this law means that the degree of disorder of a system always tends to increase. I have verified this experimentally many times through periodic observation of my office. The second law is a *macroscopic* statement; it deals with big things like steam engines, but it arises from a microscopic description of atoms and energy states provided by detailed physical theories. The laws governing these microstates are all entirely reversible with respect to time. So how can an arrow of time emerge?

Laws similar to the classical laws of thermodynamics have also been constructed to describe the properties of black holes and of gravitational fields in general. Although the definition of the entropy associated with gravitational fields is difficult to define, these laws seem to indicate that the arrow of time persists even in a collapsing Universe. It was for this reason that Hawking abandoned his time-reversal idea.

Another arrow-of-time problem emerges from quantum mechanics, which is again time-symmetric, but in which weird phenomena occur such as the collapse of the wavefunction when an experiment is performed. Wavefunctions appear to do this only in one direction of time and not the other but, as I have hinted above, this may well just be a conceptual difficulty arising from the interpretation of quantum mechanics itself.

The no-boundary hypothesis

Space and time are very different concepts to us, living as we do in a low-energy world far removed from the Big Bang. But does that mean that space and time were always different? Or in a quantum theory of gravity could they really be the same? In classical relativity theory, space-time is a four-dimensional construction wherein the three dimensions of space and one dimension of time are welded together. But space and time are not equivalent. One idea associated with quantum cosmology, developed by Hawking together with Jim Hartle, is that the characteristic signature of time may be erased when the gravitational field is very strong. The idea is based on an ingenious use of the properties of imaginary numbers. (Imaginary numbers are all multiples of the number i which is defined to be the square root of minus one.) This tinkering with the nature of time is part of *the no boundary* hypothesis of quantum cosmology due to Hartle and Hawking. Since, in this theory, time loses the characteristics that separate it from space, the concept of a beginning in time becomes meaningless. Space-times with this signature therefore have no boundary. There is no Big Bang, no singularity, because there is no time, just another direction of space.

This view of the Big Bang is one in which there is no creation, because the word creation implies some kind of 'before and after'. If there is no time then the Universe has no beginning. Asking what happened before the Big Bang is like asking what is further north than the North Pole. The question is meaningless.

I should stress that the no-boundary conjecture is not accepted by all quantum cosmologists: other ways of understanding the beginning (or lack of it) have been proposed. The Russian physicist Alexander Vilenkin proposed an alternative treatment of quantum cosmology in which there is a definite creation, through which the universe emerges by a process of quantum tunnelling out of nothing.

Theories of everything

I have tried to describe just a few of the areas in which particle physicists and cosmologists have been attempting to weld together quantum physics and gravity theory. This is one step in the direction of what many physicists feel is the ultimate goal of science: to write the mathematical laws describing all known forces of nature in the form of one equation that you might, perhaps if you have no dress-sense, wear on your T-shirt.

The laws of physics, sometimes also called the laws of nature, are the basic tools of physical science. They comprise mathematical equations that govern the behaviour of matter (in the form of elementary particles) and energy according to the various fundamental interactions described above. Sometimes experimental results obtained in the laboratory or observations of natural physical processes are used to infer mathematical rules which describe these data. Other times a theory is created first as the result of a hypothesis or physical principle which receives experimental confirmation only at a later stage. As our understanding evolves, seemingly disparate physical laws become unified in a single overarching theory. The examples given above show how influential this theme has been over the past hundred years or so.

But there are deep philosophical questions lying below the surface of all this activity. For example, what if the laws of physics were different in the early Universe? Could one still carry out this work? The answer to this is that modern physical theories actually predict that the laws of physics do change. As one goes to earlier and earlier stages in the Big Bang, for example, the nature of the electromagnetic and weak interactions changes so that they become indistinguishable at sufficiently high energies. But this change in the law is itself described by another law: the so-called electroweak theory. Perhaps this law itself is modified at scales where grand unified theories take precedence, and so on right back to the very beginning of the Universe.

Whatever the fundamental rules are, however, physicists have to assume that they apply for all times since the Big Bang. It is merely the low-energy outcomes of these fundamental rules that change with time. Making this assumption, they are able to build a coherent picture of the thermal history of the Universe which does not seem to be in major conflict with the observations. This makes this assumption reasonable, but does not prove it to be correct.

Another set of important questions revolves around the role of mathematics in physical theory. Is nature really mathematical? Or are the rules we devise merely a kind of shorthand to enable us to describe the Universe on as few pieces of paper as possible? Do we discover laws of physics or do we invent them? Is physics simply a map, or is it the territory itself?

There is also another deep issue connected with the laws of physics pertaining to the very beginning of space and time. In some versions of quantum cosmology, for example, one has to posit the existence of physical laws that exist, as it were, in advance of the physical universe they are supposed to describe. This has caused many theoreticians to adopt a philosophical approach that mirrors the ideas of Plato. In the Platonic tradition, true existence belongs to the idealized world of form rather than our imperfect world of the senses. To the Neoplatonic cosmologists, what really exists are the mathematical equations of the (yet unknown) theory of everything, rather than the physical world of matter and energy. On the other hand, not all cosmologists get carried away in this manner. To those of a more pragmatic disposition the laws of physics are simply a neat description of our Universe whose significance lies simply in their usefulness.

There have been many attempts to produce theories of everything, involving such exotic ideas as supersymmetry and string theory (or even a combination of the two known as superstring theory). In superstring theory, particles are not treated as particles at all but as oscillations in one-dimensional entities called strings. The different modes of vibration string loops correspond to different particles. The strings themselves live in a space of ten or twenty-six dimensions. Our space-time has only four dimensions (three space and one time), so the extra dimensions must be hidden. Perhaps they are wrapped up so small that they cannot be observed. After much excitement in the 1980s this idea went out of fashion, largely due to the technical complexity involved in handling such complicated multidimensional objects. More recently, these ideas have experienced a kind of renaissance, with the generalization of the concept of strings into 'branes', higher-dimensional objects whose name derives from 'membrane', and the realization that there is, in effect, a single theory (called 'M-theory') describing all versions of this kind of approach. These are exciting ideas but they are relatively undeveloped; string theory has not yet made any clear predictions that have impacted on cosmology. It remains to be seen whether the grander-than-grand unification to which these approaches aspire can actually be realized.

The search for a theory of everything also raises interesting philosophical questions. Some physicists, Hawking among them, would regard the construction of a theory of everything as being, in some sense, reading the mind of God, or at least unravelling the inner secrets of physical reality. Others simply argue that a physical theory is just a *description* of reality, rather like a map. A theory might be good for making predictions and understanding the outcomes of observation or experiment but it is no more than that. At the moment we use a different map for gravity from the one we use for electromagnetism or for weak nuclear interactions. This may be cumbersome, but it is not disastrous. A theory of everything would simply be a single map, rather than a set of different ones that one uses in different circumstances. This latter philosophy is pragmatic. We use theories for the same reasons that we use maps: because they are useful. The famous London Underground map is certainly useful, but it is not a particularly accurate representation of physical reality. Nor does it need to be.

And in any case one has to worry about the nature of explanation afforded by a theory of everything. How will it explain, for example, why the theory of everything is what it is and not some other theory? To my mind, this is the biggest problem of all. Can any theory based on quantum mechanics be complete in any sense, when quantum theory is in its nature indeterministic? Moreover, developments in mathematical logic have cast doubt on the ability of any theory to be completely self-contained. The logician Kurt Gödel has proved a theorem, known as the incompleteness theorem, that shows that any mathematical theory will always contain things that can't be proved within the theory.

The Anthropic Principles

Cosmology has always been about Man's attempts to understand the Universe and his relationship to it. As scientific cosmology has evolved, Man's role has diminished. Our existence appears accidental, unplanned, and incidental to whatever purpose the cosmos was constructed to fulfil. This interpretation has more recently been challenged by a suggestion called the Anthropic Principle, that there might, after all, be a deep connection between the existence of life and the fundamental physics that governs how the Universe evolves. It was Brandon Carter who first suggested adding the word 'Anthropic' to the usual 'Cosmological Principle' to stress the fact that our Universe is 'special', at least to the extent that it has permitted intelligent life to evolve within it.

There are many otherwise viable cosmological models that are not compatible with the observation that human observers exist. For example, we know that heavy elements like carbon and oxygen are vital to the complex chemistry required for terrestrial life to have developed. We also know that it takes around 10 billion years of stellar evolution for generations of stars to synthesize significant quantities of these elements from the primordial gas of hydrogen and helium that exists in the early stages of a Big Bang model. We know therefore that we could not inhabit a universe younger than about 10 billion years old. Since the size of the Universe is related to its age if it is expanding, this line of reasoning sheds some light on the question of why the Universe is as big as it is. It has to be big, because it has to be old if there has been time for us to evolve within it. This form of reasoning is usually called the 'Weak' Anthropic Principle and it can lead to useful insights into the properties our Universe might be expected to possess simply by virtue of our presence within it.

Some cosmologists have sought to extend the Anthropic Principle into deeper waters. While the weak version applies to physical properties of our Universe such as its age, density, or temperature, the 'Strong' Anthropic Principle is an argument about the laws of physics according to which these properties evolve. It appears that these fundamental laws are very finely tuned to permit complex chemistry, which, in turn, permits the development of biology and ultimately human life. If the laws of electromagnetism and nuclear physics were only slightly different, chemistry and biology would be impossible. On the face of it, the fact that the laws of nature do appear to be tuned in this way seems to be a coincidence, in that there is nothing in our present understanding of fundamental physics that requires the laws to be conducive to life in this way. This is therefore something we should seek to explain.

In some versions of the strong Anthropic Principle, the reasoning is essentially an argument from design: the laws of physics are as they are because they *must* be like that for life to develop. This is tantamount to requiring that the existence of life is itself a law of nature, and the more familiar laws of physics are subordinate to it. This kind of reasoning appeals to some with a religious frame of mind but its status among scientists is rightly controversial, as it suggests that the Universe was designed specifically in order to accommodate human life.

An alternative, and perhaps more scientific, construction of the strong Anthropic Principle involves the idea that our Universe may consist of an ensemble of mini-universes, each one having different laws of physics to the others. This may be what emerges from a unified theory, in which the high-energy symmetry is broken in a different way in different patches of the Universe. Obviously, we can only have evolved in one of the mini-universes compatible with the development of organic chemistry and biology, so we should not be surprised to be in one where the underlying laws of physics appear to have special properties. This provides some kind of explanation for the apparently surprising properties of the laws of nature mentioned above. This is not an argument from design, since the laws of physics could vary haphazardly from mini-universe to mini-universe.

This version of the Anthropic Principle is rightly controversial, but it at least addresses the distinction between the 'how' and the 'why'. It remains to be seen whether cosmology can explain *why* the Universe is the way it is, but we've certainly come a long way towards understanding *what* happened, and *how*.

Epilogue

Cosmology is in many ways similar to forensic science. Neither cosmologists nor forensic scientists can perform experiments that recreate past events under slightly different conditions, which is what most other scientists do. There is only one Universe, one scene of the crime. In both fields the available evidence is often circumstantial, difficult to gather, and open to ambiguity of interpretation. Despite these difficulties, the case in favour of the Big Bang is, in my view, proven beyond all reasonable doubt.

Of course, important questions remain unresolved. We still do not know the form of most of the matter in the Universe. We do not know for sure whether the Universe is finite or infinite. We do not know how the Universe began, or whether inflation happened. Nevertheless, the points of agreement between theory and observation are so many and so striking that the pieces of a coherent picture seem at last to be falling into place. But, as the saying goes, these are famous last words.

Further reading

General references

Coles, P. (ed.), *The Routledge Companion to the New Cosmology* (London: Taylor & Francis, 2001).

Gribbin, J., *Companion to the Cosmos* (London: Orion Books, 1997).

Ridpath, I. (ed.), *The Oxford Dictionary of Astronomy* (Oxford: Oxford University Press, 1997).

Chapter 1

Barrow, J. D., *The World Within the World* (Oxford: Oxford University Press, 1988).

Harrison, E., *Darkness at Night* (Cambridge, Mass.: Harvard University Press, 1987).

Hoskin, M. (ed.), *The Cambridge Illustrated History of Astronomy* (Cambridge: Cambridge University Press, 1997).

Lightman, A., *Ancient Light: Our Changing View of the Universe* (Cambridge, Mass.: Harvard University Press, 1991).

North, J., *The Fontana History of Astronomy and Cosmology* (London: Fontana, 1994).

Chapter 2

Coles, P., *Einstein and the Birth of Big Science* (Cambridge: Icon Books, 2000).

Pais, A., *'Subtle is the Lord . . . ': The Science and the Life of Albert Einstein* (Oxford: Oxford University Press, 1992).

Thorne, K. S., *Black Holes and Time Warps* (New York: Norton & Co., 1994).

Chapter 3

Eddington, A. S., *The Nature of the Physical World* (Cambridge: Cambridge University Press, 1928).

Trope, E. A., Frenkel, V. Y., and Chernin, A. D., *Alexander A. Friedmann: The Man who Made the Universe Expand* (Cambridge: Cambridge University Press, 1993).

Chapter 4

Florence, R., *The Perfect Machine: Building the Palomar Telescope* (New York: HarperCollins, 1994).

Graham-Smith, F., and Lovell, B., *Pathways to the Universe* (Cambridge: Cambridge University Press, 1988).

Hubble, E., *The Realm of the Nebulae* (Newhaven: Yale University Press, 1936).

Preston, R., *First Light: The Search for the Edge of the Universe* (New York: Random House, 1996).

Chapter 5

Barrow, J. D., and Silk, J., *The Left Hand of Creation* (New York: Basic Books, 1983).

Close, F., *The Cosmic Onion* (London: Heinemann, 1983).

Davis, P. C. W., *The Forces of Nature* (Cambridge: Cambridge University Press, 1979).

Pagels, H. R., *Perfect Symmetry* (Harmondsworth: Penguin Books, 1992).

Silk, J., *The Big Bang*, rev. and updated edn. (New York: W. H. Freeman & Co., 1989).

Weinberg, S., *The First Three Minutes* (London: Fontana, 1983).

Chapter 6

Gribbin, J., and Rees, M. J., *The Stuff of the Universe* (Harmondsworth: Penguin Books, 1995).

Guth, A. H., *The Inflationary Universe* (New York: Jonathan Cape, 1996).

Krauss, L. M., *The Fifth Essence* (New York: Basic Books, 1989).

Livio, M., *The Accelerating Universe* (New York: John Wiley & Sons, 2000).

Overbye, D., *The Lonely Hearts of the Cosmos* (New York: HarperCollins, 1991).

Rees, M. J., *Just Six Numbers* (London: Weidenfeld & Nicolson, 1999).

Riordan, M., and Schramm, D., *The Shadows of Creation* (Oxford: Oxford University Press, 1993).

Chapter 7

Chown, M., *The Afterglow of Creation* (London: Arrow Books, 1993).

Cornell, J. (ed.), *Bubbles, Voids and Bumps in Time: The New Cosmology* (Cambridge: Cambridge University Press, 1989).

Smoot, G., and Davidson, K., *Wrinkles in Time* (New York: Avon Books, 1993).

Chapter 8

Barrow, J. D., *Theories of Everything* (Oxford: Oxford University Press, 1991).

—— *Pi in the Sky* (Oxford: Oxford University Press, 1992).

—— *The Origin of the Universe* (London: Orion Books, 1995).

Coles, P., *Hawking and the Mind of God* (Cambridge: Icon Books, 2000).

Hawking, S. W., *A Brief History of Time* (New York: Bantam Books, 1988).

Lidsey, J. E., *The Bigger Bang* (Cambridge: Cambridge University Press, 2000).

Index

A

Abell clusters 95–6
accelerating universe 11, 56, 91–2
age of the Universe 54–6
Almagest 5
Alpher, R. 62
Anaximander 4
Andromeda Nebula 47, 94–5
anthropic principle 125–7
antimatter 68, 72–3
Aquinas, T. 5
Aristotle 5, 7
arrow of time 117–20

B

Babylon 2–3
baryons 64, 68, 72–3, 80; baryon
    catastrophe 83; baryon
    number 72
Bethe, H. 62
Big Bang 8–11, 37–8, 57–73, 115–17
black body 59–61
black holes 24–6, 115–16, 120
Bondi, H. 58
BOOMERANG 105–6
bosons 67–70
Brahe, T. 6, 89
branes 123

C

C-field 58
carbon 125; carbon dating 54

Cepheid variables 51–3
CERN 65
Chandrasekhar mass 90
China 2–3
classical physics 114
closed Universe 10, 34, 78, 118
clusters of galaxies 81, 95–6
COBE (Cosmic Background
    Exlorer) 98–100, 102
continuous creation 58
Coma cluster 17–18, 96
Copenhagen interpretation
    111–12
Copernican Principle 29–30,
    43–4, 81
Copernicus, N. 5, 7
cosmic microwave background
    8–9, 30, 59–61, 73
cosmological constant 31, 91–2
cosmological flatness problem,
    see flatness problem
Cosmological Principle 29–30,
    32–3, 43–4, 75, 98; Perfect
    Cosmological Principle 58–9
cosmography 93–106
curved space 20–26, 77–8

D

dark matter 11, 71, 81–5; cold dark
    matter 101–2
de Sitter, W. 8, 46
decelerating universe 54–6, 77
determinism 111–12, 114, 124
deuterium 62–3; deuterium
    bottleneck 64
Dicke R. 59

dipole anisotropy 98
Dirac, P. A. M. 67
distance indicators 51–3
Doppler effect 41–2, 98
Doppler shift, *see* Doppler effect

E

Eddington, A. S. 43
Einstein, A. 7, 11, 14–26, 28–38;
'biggest blunder' 30–1
electromagnetism 12, 15, 67, 70, 113–14
electrons 61, 68–9
electroweak theory 70–1, 113, 122
energy 75–6
entropy 119
Enuma Elish 2
equivalence principle, *see* principle of equivalence
escape velocity 76
Euclid's geometry 10, 21–2, 35
expanding Universe 8, 39–53

F

fermions 67–70
Feynman, R. 67
filaments 96
flat Universe 10–11, 33, 78;
flatness problem 86–7
fractal 96
Friedmann, A. 8; Friedmann models 32–4, 45–6, 75–8, 86, 99

G

galaxies 7–8, 26, 40, 47, 51–2, 58
galaxy clusters, *see* clusters of galaxies
Galileo 15
Gamow, G. 62
gauge bosons, *see* bosons
gedanken experiments 15–19
general relativity, *see* general theory of relativity
general theory of relativity 7–8, 20–6, 28–9, 74–5
Glashow, S. L. 70
globular clusters 56, 79
gluons 70
Gödel, K. 124
Gold, T. 58
gravitons 115
gravity 12–14, 67, 112;
gravitational instability 99;
gravitational lensing 84;
gravitational waves 115
grand unified theory (GUT) 71
Great Wall 96
Greek cosmology 4–5
GUT, *see* grand unified theory
Guth, A. 87

H

H_o, *see* Hubble's constant
hadrons 68
Hartle, J. M. 120
Hawking, S. 37, 116–18, 120
Heat Death of the Universe 6
Heisenberg, W. 108; *see also* uncertainty principle

helium 62–4, 125
Herman, R. 62
Hipparchos 51
Higgs 71
homogeneity 29, 45, 58; *see also* Cosmological Principle
Hoyle, F. 57–8
Hubble, E. 7–8, 43; Hubble's constant 48–56, 75, 78; Hubble's law 39–47
Hubble Deep Field 10
Hubble Space Telescope (HST) 10, 53, 101–2
hydrogen 62, 125

I

incompleteness theorem 124
inflation 11, 30, 73, 87–8
isotropy 29–30, 45, 58, 61; *see also* Cosmological Principle

K

Kepler, J. 6, 7; Kepler's star 89

L

Laplace 25
Large Magellanic Cloud 90
Lemaître, G. 8, 32, 43–4
leptons 67–8
Local Group 95
large-scale structure 93–106
leptons 67–9
Lick Map 93, 95
life in the Universe 125–7
lithium 62

M

M theory 123
Mach's Principle 29
Many Worlds interpretation 112
MAP (Microwave Anisotropy Probe) 10, 84, 104
Marduk 2–3
MAXIMA 105
Maxwell, J. C. 15; Maxwell's theory 15, 64, 70
Michell, J. 25
Milky Way 7–8, 30, 50, 52, 89
mythology 1–4

N

Narlikar, J. V. N. 58
nebulae, *see* galaxies
neutrinos 67–8
neutrons 62–4
Newton, I. 6
Newtonian mechanics 13, 27–8, 110
no-boundary hypothesis 120–1
nucleosynthesis 62–4, 80

O

Olbers' Paradox 7
Omega (Ω) 74–92
open Universe 10, 34, 56, 78
oxygen 125

P

Pan Gu 2–3
parallax 50–1

parallel universes, *see* Many
    Worlds interpretation
Peebles, P. J. E. 59
Penrose, R. 37, 116–17
Penzias, A. 8, 59
phase transitions 11
photons 67–70, 108–11
Planck, M. 107; Planck time 115;
    117; Planck length 115
Planck Surveyor 10, 84, 104
Plato 4–5; Platonism in
    cosmology 6, 123
Principia 6
principle of equivalence
    17–19
principle of relativity 7–8
protons 62–4
Ptolemy 5

Q
QCD 69–70
QED 67–70, 114
quantum chromodynamics, *see*
    QCD
quantum electrodynamics, *see*
    QED
quantum physics 107–14
quantum gravity 66, 88, 112–14,
    117
quarks 68–70, 88
quasars 66

R
radioactivity 54–5
recombination 61
redshift 42–7, 90, 94–5, 102

relativity, general theory of,
    see general theory of
    relativity
relativity, principle of, *see*
    principle of relativity
relativity, special theory of, *see*
    special theory of relativity
ripples 10, 98–100, 102–4
Ryle, M. 58

S
Sakharov, A. 72–3
Salaam, A. 70
Sandage, A. R. 79
Schrödinger, E. 108; Schrödinger
    equation 109; Schrödinger's
    cat 111
Schwarzschild, K. 25–6;
    Schwarzschild radius 26;
    Schwarzschild solution 26, 36,
    115
Shapley concentration 96
singularity 9, 34, 36–7, 115–17;
    naked 37
Slipher, V. 40
space, curved, *see* curved space
space-time 8, 10, 32, 117–20;
    'foam' 115–16
special relativity, *see* special
    theory of relativity
special theory of relativity
    14–17
spectroscopy 40–3
static universe 30–1
standard model of particle
    physics 71

Steady State 57–9
string theory 57
strong anthropic principle 125
strong energy condition 117
strong nuclear force 12, 70, 113
superclusters 97–8
supernovae 51, 89–91
superstrings 123
symmetry 28–9, 70, 117;
    supersymmetry 71, 123

T
Thales 4
thermodynamics 6; second law
    119
theories of everything 121–5
Tiamat 2–3
time 117–18
two-degree field survey 94,
    97

U
uncertainty principle 108–14

V
vacuum energy 92, 106; see also
    cosmological constant
Vilenkin, A. 121
Virgo cluster 53, 96
voids 96–8

W
wavefunction 108–11
weak anthropic principle, see
    anthropic principle
weak nuclear force 12, 69–70, 113
Weinberg, S. 70
Wilson, R. 8, 59

X
X-rays 83

Expand your collection of
VERY SHORT INTRODUCTIONS

1. Classics
2. Music
3. Buddhism
4. Literary Theory
5. Hinduism
6. Psychology
7. Islam
8. Politics
9. Theology
10. Archaeology
11. Judaism
12. Sociology
13. The Koran
14. The Bible
15. Social and Cultural Anthropology
16. History
17. Roman Britain
18. The Anglo-Saxon Age
19. Medieval Britain
20. The Tudors
21. Stuart Britain
22. Eighteenth-Century Britain
23. Nineteenth-Century Britain
24. Twentieth-Century Britain
25. Heidegger
26. Ancient Philosophy
27. Socrates
28. Marx
29. Logic
30. Descartes
31. Machiavelli
32. Aristotle
33. Hume
34. Nietzsche
35. Darwin
36. The European Union
37. Gandhi
38. Augustine
39. Intelligence
40. Jung
41. Buddha
42. Paul
43. Continental Philosophy
44. Galileo
45. Freud
46. Wittgenstein
47. Indian Philosophy
48. Rousseau
49. Hegel
50. Kant
51. Cosmology
52. Drugs
53. Russian Literature
54. The French Revolution
55. Philosophy
56. Barthes
57. Animal Rights
58. Kierkegaard
59. Russell
60. Shakespeare
61. Clausewitz
62. Schopenhauer
63. The Russian Revolution
64. Hobbes
65. World Music
66. Mathematics
67. Philosophy of Science
68. Cryptography
69. Quantum Theory
70. Spinoza

71. Choice Theory
72. Architecture
73. Poststructuralism
74. Postmodernism
75. Democracy
76. Empire
77. Fascism
78. Terrorism
79. Plato
80. Ethics
81. Emotion
82. Northern Ireland
83. Art Theory
84. Locke
85. Modern Ireland
86. Globalization
87. Cold War
88. The History of Astronomy
89. Schizophrenia
90. The Earth
91. Engels
92. British Politics
93. Linguistics
94. The Celts
95. Ideology
96. Prehistory
97. Political Philosophy
98. Postcolonialism
99. Atheism
100. Evolution
101. Molecules
102. Art History
103. Presocratic Philosophy
104. The Elements
105. Dada and Surrealism
106. Egyptian Myth
107. Christian Art

Visit the
VERY SHORT INTRODUCTIONS
Web site

www.oup.co.uk/vsi

➤ **Information** about all published titles

➤ News of **forthcoming books**

➤ **Extracts** from the books, including titles not yet published

➤ **Reviews** and views

➤ **Links** to other **web sites** and main OUP web page

➤ Information about **VSIs in translation**

➤ **Contact** the editors

➤ **Order** other **VSIs** on-line

GALILEO
A Very Short Introduction
Stillman Drake

Galileo's scientific method was of overwhelming significance for the development of modern physics, and led to a final parting of the ways between science and philosophy.

In a startling reinterpretation of the evidence, Stillman Drake advances the hypothesis that Galileo's trial and condemnation by the Inquisition in 1633 was caused not by his defiance of the Church, but by the hostility of contemporary philosophers.

Galileo's own beautifully lucid arguments are used to show how his scientific method was utterly divorced from the Aristotelian approach to physics in that it was based on a search not for causes but for laws.

'stimulating and very convincing'

Theology

www.oup.co.uk/isbn/0-19-285456-9

INTELLIGENCE
A Very Short Introduction
Ian J. Deary

Ian J. Deary takes readers with no knowledge about the science of human intelligence to a stage where they can make informed judgements about some of the key questions about human mental activities. He discusses different types of intelligence, and what we know about how genes and the environment combine to cause these differences; he addresses their biological basis, and whether intelligence declines or increases as we grow older. He charts the discoveries that psychologists have made about how and why we vary in important aspects of our thinking powers.

'There has been no short, up to date and accurate book on the science of intelligence for many years now. This is that missing book. Deary's informal, story-telling style will engage readers, but it does not in any way compromise the scientific seriousness of the book . . . excellent.'

Linda Gottfredson, University of Delaware

'Ian Deary is a world-class leader in research on intelligence and he has written a world-class introduction to the field . . . This is a marvellous introduction to an exciting area of research.'

Robert Plomin, University of London

www.oup.co.uk/isbn/0-19-289321-1

POSTMODERNISM
A Very Short Introduction
Christopher Butler

Postmodernism has become the buzzword of contemporary society over the last decade. But how can it be defined? In this Very Short Introduction Christopher Butler lithely challenges and explores the key ideas of postmodernism, and their engagement with literature, the visual arts, film, architecture, and music. He treats artists, intellectuals, critics, and social scientists 'as if they were all members of a loosely constituted and quarrelsome political party' – a party which includes such members as Jacques Derrida, Salman Rushdie, Thomas Pynchon, David Bowie, and Micheal Craig-Martin – creating a vastly entertaining framework in which to unravel the mysteries of the 'postmodern condition', from the politicizing of museum culture to the cult of the politically correct.

> 'a preeminently sane, lucid, and concise statement about the central issues, the key examples, and the notorious derilections of postmodernism. I feel a fresh wind blowing away the miasma coiling around the topic.'
>
> **Ihab Hassan, University of Wisconsin, Milwaukee**

www.oup.co.uk/isbn/0-19-280239-9